TIME IS AN ARTIST

Angkor, late 12th century. 1963

ACKNOWLEDGEMENTS

My wife, Hazel, accompanied me on the long trips on which most of the photographs were taken. Taking the pictures, I usually walked alone; but it made a difference to know that I was not totally alone.

Moshe Barasch called my attention to some very interesting books and articles.

Getting a work with so many color photographs published required a long uphill fight. Nicolas Ducrot bore the brunt of it, gave me unswerving support, and prevailed.

In the fall of 1977, Jill Anderson helped me to improve the text. Her total reliability and exceptional sensitivity made it delightful to work with her.

Jo Kaufmann did a splendid job copyediting the text and made many helpful suggestions.

Books by Walter Kaufmann

NIETZSCHE: PHILOSOPHER, PSYCHOLOGIST, ANTICHRIST
CRITIQUE OF RELIGION AND PHILOSOPHY
FROM SHAKESPEARE TO EXISTENTIALISM
THE FAITH OF A HERETIC
HEGEL
TRAGEDY AND PHILOSOPHY
WITHOUT GUILT AND JUSTICE
EXISTENTIALISM, RELIGION, AND DEATH
THE FUTURE OF THE HUMANITIES

Photographs and Text

RELIGIONS IN FOUR DIMENSIONS
MAN'S LOT: A TRILOGY
LIFE AT THE LIMITS
TIME IS AN ARTIST
WHAT IS MAN?

Verse

GOETHE'S FAUST: A NEW TRANSLATION
CAIN AND OTHER POEMS
TWENTY-FIVE GERMAN POETS

Translated and Edited

EXISTENTIALISM FROM DOSTOEVSKY TO SARTRE
JUDAISM AND CHRISTIANITY: ESSAYS BY LEO BAECK
PHILOSOPHIC CLASSICS, 2 volumes
RELIGION FROM TOLSTOY TO CAMUS
HEGEL: TEXTS AND COMMENTARY
HEGEL'S POLITICAL PHILOSOPHY
MARTIN BUBER'S I AND THOU

Nietzsche Translations

THE PORTABLE NIETZSCHE
(Thus Spoke Zarathustra, Twilight of the Idols,
The Antichrist and Nietzsche contra Wagner)
BASIC WRITINGS OF NIETZSCHE
(The Birth of Tragedy, Beyond Good and Evil,
On the Genealogy of Morals, The Case of Wagner, Ecce Homo)
THE WILL TO POWER
THE GAY SCIENCE

Library of Congress Cataloging in Publications data

Kaufmann, Walter Arnold
  Time Is an Artist

  Bibliography: p.
  1. Photography,Artistic 2. Time. I. Title
TR654.K38   779'.092'4   78-9074
ISBN 0-07-033317-3

This book was originated and produced by
VISUAL BOOKS, INC.
342 Madison Avenue
New York, New York 10017

TIME IS AN ARTIST

Photographs and Text
by
WALTER KAUFMANN

Princeton. 1949

READER'S DIGEST PRESS
Distributed
by
McGraw-Hill, Inc.
NEW YORK
1978

Ein Grabmal für

BRUNO KAUFMANN
1881–1956
EDITH KAUFMANN
1887–1977

Near Carmel, California. 1958

Olympic Peninsula, Washington

Balaam

Slandered for centuries in verse
by lovers while the friends of art
blindly gave her the devil's part,
she waits for one who, called to curse,
might look and then refuse to chime
in, as he saw, surprised,
her stunning works and recognized
the greatest artist of all—time.

CONTENTS

NOTES ON THE COLOR PHOTOGRAPHS

NOTES ON THE COLOR PHOTOGRAPHS
Time's Artistry in Nature: 1–25

1 Grand Canyon of the Colorado River, Arizona
Skies
2 Villa Serbelloni, Belaggio, Italy
3 Kauai, Hawaii
4 Florida coast near Orlando
5 Uluwatu temple, Bali
6 Canberra, Australia
7 Louisiana
Rocks and mountains
8 Garden of the Gods, Colorado
9 Yellowstone National Park, Wyoming
10 Monument Valley, Utah
11 Gung Batur, Bali
12 Dolomites, South Tirol, Italy

13 Pacific coast, Olympic Peninsula, Washington
14 Pacific coast, Bandon, Oregon
Trees and colors
15 Tree stump with fungi, Princeton, New Jersey
16 Flowers, King's Park, Perth, Australia
17–18 Fall in Princeton
19 Ormiston, outback, Australia
20 Bark, Canberra
21 Olive tree, Israel
22 Bark, Arizona
23 Inside of a tree, rainforest, Olympic Peninsula
24 Canberra
25 Nature reclaiming a wooden door, Soglio, Switzerland

Time's Work on Art: 26–72

Patina (from the author's collection)
26 Syrian glass, 2nd century A.D. (12 cm and 9.5 cm)
27 Bronze Avalokiteshwara, Khmer, Bayon style, ca. 1200. (19.5 cm)
28 Bronze Buddha, Burma, 18th century. (18 cm)
29 Bronze Buddha, Chiengsen, Thailand, 14th century. (12 cm)
30 Bronze stag, Iran, 500 B.C. (15x17.5 cm)
Reflections: changes wrought by light and water
31 Light softening bronze Buddhas without patina, Bangkok
32 Canal near Bangkok: the artistry of change
33 Isola Bella, Italy: gently moving water improving bad art
Walls
34–35 Villa Serbelloni, Belaggio, Italy
36 Venice
37 Bali
Ruins and restoration
38–39–40 Three views of the old Indian ruins, Mesa Verde, Colorado, 1969
41–42 Prambanan, Java, 1974
43 Sewu, near Prambanan, Java, 1974
44 Angkor, 1963
45–49 Borobodur, Java
45 Restoration in progress: before and after side by side, 1974
46 Before the cleaning, 1971
47–49 After the cleaning; 49 shows the end result, 1974
The torso as an illustration of time's artistry
50 Torso of the *Venus of Cyrene*, Roman copy of a 4th century Greek statue, Terme Museum, Rome. (1.52 m)
51–52 The widely admired *Venus* of the Capitoline Museum, which is a Roman copy of a Hellenistic original of the 3rd century B.C., and a fragmentary Hellenistic original in the Palazzo

dei Conservatori. How much could each be improved by amputations?
Restoration, patina, and light
53 "Repair" at the Taj Mahal, Agra, India, 1975. It is often said that restorers do not destroy anything.
54 Greek temple at Paestum, Italy, 5th century, with patina, 1970.
55 Detail of one of the two bronze moors on the Clock Tower, Piazza San Marco, Venice. They "have been striking the hour for 500 years"— and time has rewarded them with a rich green patina. 1975
56 The effects of passing time in San Pietro, Assisi, Italy, 1976
57 The effects of light in the Taj Mahal, 1975
Patina and restoration in Jerusalem
58–59–62 The Dome of the Chain, still unrestored in 1975
60-61 Small unrestored structures with patina; in 61 the restored Dome of the Rock in the background, 1975
63 Middle Bronze Age pot found near Jerusalem, ca. 2000 B.C., with patina. (26 cm) Author's collection
64-65 Dome of the Rock after restoration, 1975
Al Aqsa Mosque, 1975
66 South wall. Israel was excluded from UNESCO for the dig that exposed the old foundations that go back to New Testament times. The silver covering of the dome is part of the latest Muslim restoration of the mosque.
67-68 Restoration in progress: the old on the left and the new on the right
69 Discarded Byzantine capital with patina
70 Church of the Holy Sepulcher, 1975: a corner on the outside not yet reached by restorers
71 Brand new pillars without patina
72 Egyptian mummy, Vatican Museum, 1977

1

4

5

6

7

9

10

1

2

15

16

19

20

I

TIME DISCOVERED OR TIME IN TEMPORAL PERSPECTIVE

1

Time is an artist. But an artist is not only an artist. An artist may also be kind, cruel, or savage. Nor do all artists make pretty things. Michelangelo's contemporaries were awed by his *terribilità*. Rembrandt's contemporaries did not find his greatest works beautiful, and Turner's contemporaries hardly considered his masterpieces works of art. Goya's most original creations still strike terror into our hearts. Time is an artist, but it is more like these men than like Giotto, Fra Angelico, or Botticelli.

The Grand Canyon is the work of time, done in collaboration with the Colorado River. The Garden of the Gods near Colorado Springs, Monument Valley in Navaho land in Utah, and the Dolomites in the Alps are among time's works no less than the rocky Oregon coast and Bali's volcanoes. Nature abounds in wonders wrought by time.

One can think of art as a bold attempt to conquer time, as setting its face against change and decay. Yet ancient glass, buried, becomes iridescent; and bronze statues, too, are transfigured by time and made much more beautiful than they were in their prime. Time paints the foliage of fall, evening skies, old trees, and many things made by man.

Time is often destructive—as sculptors are when they work on a piece of stone. Yet old faces can be much more expressive than young ones, old walls and sculptures much richer than new ones.

What matters here is not to prove a thesis, but to oppose one sensibility to another; to lead people to see and feel differently; to call into question one way of experiencing life and to develop another. Attitudes toward time are pervasive. They involve attitudes toward the old and the young; the ancient and the new; past, present, and future; history; life and death; and ourselves. For a large part of their lives, women have a special relation to time, and attitudes toward time also influence the feelings of men and women toward women. If we would understand the human lot, we must contemplate time, for it is our lot to exist in time.

This is another way of exploring life at the limits, of seeing life as a movement beyond former limits, passing beyond them like clouds changing colors and shapes in a spectacular sunset. We always live at the limits, but we are rarely aware of it. It is the aim of this essay and these photographs to make us aware of our condition and to invite critical reflection on the attitudes mentioned, including those toward women, decay, and restoration.

It is always tempting to explore one detail exhaustively because that is safer, and it is dangerous to speak of many things briefly. But it is at least worth trying to see together what is usually studied separately, and anyone attempting to fathom man's lot can do nothing less. To go to one's death without having thought about man's lot is to die like a dog.

2

Time is unlike space. Space is no artist. Space does not *do* anything. We do not point to the works of space. Although we frequently couple time and space, they are different in these and other ways.

Time invites personification; space does not. Time is experienced as a threat, an enemy, a healer of wounds. We may count on her, or quixotically try to oppose her. We may think of her as irresistible, as the inexorable *par excellence*. No oration, entreaty, or prayer can stay her relentless course, or move her back, however slightly. None of this imagery is applied to space.

Architects, sculptors, and painters are keenly aware of space, but not as a threat, an opponent, an irresistible force. With time they must reckon in a totally different way.

Since the second World War, of course, people have spoken of conquering space and exploring space, but what is meant is space in another sense, and not what we might call spatiality.

So-called space programs are designed to explore specific places or areas between the earth and the moon or some planet. Space in the sense in which it is often paired with time is not the object. Men do not fight the effects of space; they do not even think of space as having effects the way time does.

Vast sums are spent by modern man to do battle against the effects of time. Programs of this sort are aptly called restoration. The aim is to restore things made by man to a prior state by undoing the work of time. Time is seen as the enemy, and restorers attempt to turn back the clock. This refusal to accept the world as it is, and this ambivalence toward what is old—interest in it coupled with a horror of what looks old, and the feeling that only what looks new is beautiful—that is modern humanity. What, then, is modern man? *Nouveau riche.*

He would like to be cultured and knows this involves respect for some old things, but his taste is vulgar and he likes what is gaudy, quickly accessible like a prostitute. In print he is accustomed to journalism with flashy headlines and opening paragraphs that seem to reveal the whole story. He has no time and no feeling for time and her texture. He lives in the present—he thinks—and wonders why he has little time for pleasure. In fact, he has no time for the present because he lives in the future, in worries and dreams, but mostly in worries. His perception of time is warped because, without knowing it, he has become her slave; for he has never raised himself up to face her and learn to live with her.

A good life is lived in friendship with time, seeing her as she is, respecting her works—including decay and death—coming to terms with age, with the past, and with history. Restoration is self-deception and often a crime. But to see that, we must first go back in time and see how man discovered her.

3

Most people never think about the nature of time. In antiquity hardly anyone did. One was aware, as many animals are, of the differences between day and night and of the seasons; one could not help being aware of change, but one did not ask, What is time?

Time is the dimension of change. Without awareness of change, there is and can be no awareness of time. And different attitudes toward time are corollaries of different attitudes toward change. Change is more concrete, time more abstract. The concept of time is born of reflection on change; more specifically, of attempts to measure change. These involve bold feats of abstraction. Change is spatialized: one posits points, draws imaginary lines from one point to another, and then divides these lines into segments which are thought of as periods of time.

The King James Bible shows this very well in its rendering of a number of passages in which the Hebrew original employs different phrases. Here is the very first occurrence of "space" in the King James Bible, Genesis 29.14: "And he abode with him the space of a month." A marginal gloss informs the reader, correctly, that the Hebrew text has "a month of days." In Leviticus 25.8, the King James says: "and the space of the seven sabbaths of years shall be unto thee forty and nine years." Here the Hebrew has "days" instead of "space." In the "Authorized" English version of the Five Books of Moses, "space" occurs only four times, and the last instance is in verse 30 of the same chapter: "And if it be not redeemed within the space of a full year . . ." Again, the original does not speak of space. The ancient Hebrews' attitude toward time revolutionized civilization.

Western man's attitude toward the so-called Old Testament is strangely mixed up. Countless Hebrew names, metaphors, verses, poems, stories, ideas, and attitudes have entered the languages and the cultures of Western countries, and are not only familiar but widely taken for granted, without any sense that they come from a distant place and time, and that in some cases they have been modified in the course of transmission. But genuine knowledge of the Bible has become rare. Even at the best universities, most students can no longer name five Hebrew prophets, though one need not look beyond the table of contents to find at least fifteen; and scarcely anyone has any notion of what was distinctive in ancient Hebrew attitudes—toward change, for example, or time.

The best way to discover that, and to find out how they slowly changed human thought in much of the world, is to begin, not with them, but if possible with their polar opposites. Is there a high civilization that can be viewed without undue distortion as the antithesis of the outlook of Moses and the prophets? What is at issue is not a merely antiquarian interest in how some people felt thousands of years ago, but rather, a clear grasp of fundamental alternatives. Sometimes characters in plays, poems, or novels show us impressive alternatives, but for the most part the imagination of even the greatest poets is circumscribed by their own culture and does not produce characters whose whole outlook is *radically* different. Most of the great philosophers have even less imagination than the great poets.

In his *Poetics* (9:51b), Aristotle said: "A poet differs from a historian, not because one writes verse and the other prose—the work of Herodotus could be put into verse, but it would still remain a history, whether in verse or prose—but because the historian relates what happened, the poet what might happen. That is why poetry is more philosophical and nobler than history; poetry deals with general truths, history with specific events."

Aristotle's erudition and intelligence are beyond question. And we can safely agree that Aeschylus, Sophocles, and Euripides wrote tragedies that were more philosophical and nobler than Herodotus' history. But even they could not do what history now enables us to do: to explore radically different attitudes toward time, change, and history itself.

4

Those who wished to compare ancient Israel with a profound alternative found it in Greece. Outstanding examples include Leo Baeck's "Two World Views Compared" in *The Pharisees and Other Essays*—to name only one of his efforts in this vein—and Hans Kohn's contrast of "Israel and Hellas" in Chapter II of *The Idea of Nationalism*. Both studies are very learned and thoroughly worthwhile. It is important to understand how different the two major sources of Western civilization were. Short of that, one does not really understand either them or our own frequently inconsistent ideas, which are due in part to this dual heritage.

The trouble with those essays is that the Greeks are misrepresented in order to sharpen the contrast. A single sentence from Baeck's suggestive essays shows this quite well: ". . . the Eleatic-Platonic-Aristotelian philosophy . . . is the true Greek philosophy; for Heraclitus played the role of the devil's advocate, presenting his arguments only to be refuted, and Democritus never really played any part in this philosophy" (p. 128).

To be sure, the influence of Heraclitus and Democritus did not match that of Plato and Aristotle, and Plato does approximate the antithesis of the ancient Hebrew outlook, though even he did not go all the way. But to see Plato, the radical critic of his own culture, as its most typical representative is a mistake. Nor can generalizations about the Greeks be based largely on Greek philosophy to the exclusion of the great poets. Kohn devotes 35 pages to Israel and Hellas without once mentioning Homer, Aeschylus, or Sophocles, and his two passing references to Euripides do not even come close to asking how the outlook of this great poet compared with that of the ancient Hebrews.

Plato, an Athenian, wrote his dialogues after Athens had been defeated in a great war that had lasted almost thirty years. As he wrote, her glory was but a memory; all her greatest poets, philosophers, architects, and historians were dead, and he was the last survivor with gifts to match theirs. To say that he wrote in an age of decline would be, from his own point of view, an understatement. He was the survivor of a cataclysmic disaster that had destroyed a center of civilization which had long come to see itself as *the* center of civilization. Within five years, Euripides and Sophocles had died, Athens had lost the war, Thucydides had died, and Socrates had been put to death. Plato's attitudes toward time and change were affected profoundly by these events and cannot be seen as typically Greek. Moreover, Plato's philosophy was shaped decisively by the conflicting influences of Socrates and Pythagoras, the Eleatic school of philosophy and the great poets. Nor were all the crucial influences on his thought Greek. Precisely his views of change and time drew heavily on another culture: India.

5

Ancient India is the antithesis of ancient Israel, and it is here that we find the earliest roots of Western philosophy. Indian influence on Greek thought began before Plato and is palpable in Pythagoras and the Eleatic school. The latter derives its name from Elea in southern Italy and may have been influenced by the Pythagoreans who lived nearby.

One of the unwritten rules for scholars concerned with Greek philosophy is to ignore non-Greek sources, including even Persia, not to speak of India. This is rather odd, considering that Greek philosophy began in Asia Minor, in the Persian empire; that the Persians invaded Greece in 490 and again in 480; that Herodotus, roughly 55 years older than Plato, visited Egypt and Mesopotamia; and that parts of India belonged to the Persian empire. Nobody would deny the crucial Egyptian influence on the beginnings of Greek sculpture and architecture, but most historians of Greek philosophy and Plato scholars treat the non-Greek world as if it had not existed. This is a weird example of academic departmentalization.

The notion that time, change, and multiplicity are unreal, and that ultimate reality is one, unchanging, and timeless was born in India, like the belief in transmigration, which Plato adopted from Pythagoras. These ideas are found in the Upanishads before they appear among the Greeks, and there are no good reasons for supposing that the similarities between Plato's class system in the *Republic* and the Hindu caste system are mere coincidences. The Eleatic and Platonic philosophies were not primordially Greek; the idea that perfection is timeless and change is a blemish was an import from India; and in important respects Plato represents a break in Greek culture.

Most philosophers do not care where ideas came from, but only whether attempts to prove them are valid. The paradoxes of Zeno of Elea, which were meant to prove that time, motion, and change must be illusory, have elicited a vast literature, and philosophers are still writing about them, while the human implications of such a view and its cultural ramifications have gone unexplored.

Plato certainly did not succeed in proving that the world of change is mere appearance and that ultimate reality is timeless. He differed from the school of Elea and the Upanishads by not accepting their doctrine that ultimate reality is one and that multiplicity is also merely apparent. Although he owed much to several of his predecessors, none of them shared his world view, which was distinctive and emphatically not *the* Greek view.

If our concern were only with the logical implications of the claim that change and time are unreal, we could dispense altogether with historical illustrations. It would suffice to point out what is obvious anyway: that survival requires us to act as if change were real, that the claim simply cannot be taken seriously in practice, and that any attempt to take it so has to involve some distinction between degrees of reality. One must allow change and time some reality, and one has to be content to claim that in the last analysis something timeless is in some sense "more real."

It is not surprising that philosophers should find it interesting to determine just how some philosopher of the past made the needful distinctions and to study the solutions proposed by Plato or by Shankara, who was the greatest Hindu philosopher. But if our concern is with the actual psychological and cultural ramifications of the belief that change is not real, Greece fails us because Plato's belief was never shared widely. By far the best place to discover what we are after is India.

6

In books about India or Indian philosophy one looks in vain for an answer to the question of how the view that time is unreal developed, or when it originated. Part of the problem is that people in India have not taken time seriously and have taken no interest in historical questions. Although many Indians assume to this day that their culture is the oldest anywhere because it goes back to times immemorial, the first man in India who left us some written records was the emperor Ashoka, who converted to Buddhism in the third century B.C., more than half a century after the invasion of Alexander the Great. And the only individual in India before him who can be dated with a high probability is the Buddha, who seems to have lived from 563 to 483.

It is not only the dates that pose a problem. There are virtually no other human beings in Indian antiquity who confront us as individuals. While the Old Testament is full of sharply drawn characters who can be dated, the Indians obviously took little or no interest in individuals. It makes sense to ask whether they had the concept of the individual.

Nevertheless, the origin of the idea that time is unreal can be reconstructed. It was most likely derived from an experience induced by a drug. In the Vedic hymns this drug is mentioned frequently, and one whole book of the Rigveda consists of hymns addressed to it. We know its name, Soma juice, but not what precisely it was. The juice was associated with the great god Indra, who was actually "known as the Soma drinker, armed with thunder" (X:119). In a sacrificial rite, Soma juice was offered to Indra, and some of his worshipers enjoyed it, too. Now I assume that the state produced by the drug was felt to be godlike and superior to common consciousness, and that it involved oblivion of time. This came to be considered proof that time was unreal and, as it were, a hallucination that went with inferior forms of consciousness.

The question was which state of consciousness was higher and normative: the god's or ours. Obviously that of the god and his priests, the Brahmins. At least that was the Brahmins' doctrine.

We encounter this claim in the Upanishads, but without any reference to Soma or other drugs. By that time the Brahmins had developed techniques for obtaining such states without drugs. How many of them were able to reach such mystical states or how often they reached them we do not know. They committed nothing to writing and seem to have been illiterate, and in the Upanishads, which were long transmitted orally, they were very secretive about their teachings. But what they talked about at enormous length was a state revealing the one, unchanging, timeless reality which they called Brahman or Atman. Roughly speaking, Brahman is the world soul, and Atman our own soul. Etymologically, "Atman" is related to the German *atmen*, which means to breathe, and breath is also one of the meanings of the Latin *spiritus*. The Brahmins taught that Brahman and Atman were one, and the sole ultimate reality.

The Upanishads (literally, the seminars) consist of questions the Brahmins were asked, and of their answers. The questioners lived with the priests and waited on them, hoping to acquire some of their secret wisdom, not in a day or two, but eventually. Time being considered unreal, we naturally do not know how long the instruction lasted or how many attained the experience of Brahman-Atman.

7

The enormous importance of the Upanishads lies in the fact that the intellectual elite of India denied that time was real, and that the Brahmins, who were priests and teachers, had no competition from prophets or secular teachers. The one group that might have taken an interest in history and in the deeds and the characters of outstanding individuals did not do so. Nor did they take an interest in art. Nothing individual, nothing particular was of central concern to them. They lacked the concept of the unique—of unique events, of unique acts or works, of the uniqueness of a human being. What was ephemeral did not matter; what was one among others was not real.

Suffering was unreal. The quest for a higher consciousness precluded attention to everything transitory, including persons and pain. For Moses and the prophets, religion required attention to the needs of orphans, widows, and strangers. For the sages of the Upanishads the religious quest required that one should not be distracted by the lot of the unfortunate and the oppressed.

The society of the Upanishads was a caste society. The Buddha rejected the caste system along with the authority of the Vedas; none of the Brahmins did. They never expressed any compassion for slaves or outcastes, orphans, widows, or strangers. In modern times, the British finally put an end to the Indian practice of burning widows on their husbands' funeral pyres.

The Buddha taught that the Brahmins' questions about Atman and Brahman did not tend "toward edification." He sent out missionaries, and later Ashoka sent missionaries to many foreign countries. In Sri Lanka, Burma, Thailand, and the Far East Buddhism prevailed. In India the Brahmins prevailed, and in the twentieth century Gandhi's epic battle against the caste system was to remain largely unsuccessful. But are any of these facts related to the denial of time and change?

The faith that perfection precludes change militates against all reform. The Vedas and the caste system were believed to be timeless and therefore older by far than any non-Indian culture. They had no history and had to be preserved. Differences between people were fixed eternally and were not merely products of this war or that misfortune. A slave was a slave and not an unfortunate human being. There was nothing like the Mosaic refrain that you should remember that you were slaves in the land of Egypt, and that you should know how the slaves and the stranger feel.

In some ways Hinduism changed very drastically—although few Hindus know it, for a sense of history is still very rare in India. During the centuries when it competed with Buddhism, Hinduism absorbed the old native cults that antedated the invasion of the Aryans whose religion we find in the Vedas and the Upanishads. Hinduism adopted the worship of Shiva, who is not mentioned in the Vedas and the Upanishads; the phallic worship of the lingam that is associated with Shiva; the use of idols and temples, of which there is no trace in the Vedas and Upanishads; and a profusion of gods and goddesses not found in the old traditions either. But the very idea that the use of temples and idols might have an origin in time and history is foreign to Hinduism. It is assumed that everything in one's religion has always been as it is now.

8

One Hindu idea that is first encountered in the Upanishads could have mitigated the inhumane aspects of Hinduism: the doctrine of transmigration. This involves time and change, some continuity between man and man, male and female, humanity and the animal kingdom. But this doctrine was coupled with the belief in the law of karma, which governs rebirth. After death every being gets the life it deserves. The outcaste merits his wretched lot, the slave his, and the widow hers.

Men and women have duties, but there are no human rights. Nor is respect due every human being simply because he or she is human. Cows are allowed in the temple compounds, indeed cow dung is plentiful in them, but an outcaste child who entered through the gate to retrieve a ball was beaten to death by a temple priest as recently as 1975.

Still, it may be felt that in one sense the Hindus respect the works of time. Do they not revere what is old? Do not many educated Indians still boast that their Vedas are at least 10,000 years old? They do, but the very conception of age is alien to them. They simply cannot imagine that the Vedas ever did not exist, and they refuse to think how old precisely the Vedas might be and when and where they orginated.

They think of their art as terribly ancient, though in fact not one building or sculpture unearthed in India antedates the third century B.C., and almost all of the marvels of Indian art are medieval. (The Indus Valley culture unearthed in Pakistan is pre-Vedic and came to an end in the sixteenth century, before the Aryan invasion.) In the 1960s the Prince of Wales Museum in Bombay still sold off medieval sculptures for five, ten, or fifteen dollars. The most widely admired temple in India was the Baroque Meenakshi temple in Madurai, which was begun in the seventeenth century. After India had gained her independence, this temple was restored and its myriad sculptures repainted in gaudy colors.

The superb remnants of the great sun temple at Konarak, which bear witness of the passage of time, stir no remotely comparable enthusiasm in India and attract tourists, Indian as well as foreign, merely as a curiosity. People come to gawk at the erotic carvings. But one can hardly compare Madurai, which is a place of worship, with Konarak, which is in a way more like the Parthenon or the Pyramids. For a site of that kind, a historical monument that bears witness to a past age, few Indians have any sense. In Israel, which still is the polar opposite, archaeology is a national obsession.

The oldest great art in India is Buddhist: the stupas at Sanchi with their magnificent carved gates, and the cave temples at Ajanta. Visitors to both include Hindus carrying transistor radios and "doing" these places at a speed exceeding that which is widely associated with Americans.

9

Many of the points stressed here are not peculiar to India. It is Israel that is, and for more than three thousand years has been, the exception. India merely embodies the opposite attitudes in the most striking, impressive, and profound way, while Greece, as we shall see, is in some ways a halfway house.

Certainly India is not unique in having no living relationship to its own past. In Greece and Egypt, Mesopotamia and Iran, Cambodia and Java, it also remained for foreigners to unearth and begin the study of the ancient art and literature. Only the Jews and the Chinese have kept studying their own traditions, and in both cases the reason was the same: an unusual attitude toward time and history.

A light-hearted poem in three parts may illustrate different attitudes toward time very succinctly:

1. Javanese Store

This is very, very old.
    How old?
About forty years.
    (Where does that leave me?)

This is seventeen six
one seven six
from the palace in Solo
seventeenth century
about one hundred fifty years old.

2. Varanasi Shop

This is Buddha
very old
he said
and reading the label
fifth century A.D.

    But this head is not old
    said I
    it is new.

No
it is old
sir
very old
more than five years.

You want to see really old things
come to my house
to see really old things
a thousand
two thousand
ten thousand years old.

3. Near Bethlehem

This is not old
it was built by the TURKISH Solomon
four hundred fifty years ago.

10

It is often said that the Greeks were the first to write history, and that Herodotus, in the fifth century B.C., was "the father of history." Yet the Greeks had very little historical sense. They knew next to nothing of their own history before the fifth century, and as a result we do not know much about it either. We have a few quotations from the works of the earliest Greek philosophers, who lived in Asia Minor in the sixth century, and a few remarks about them, written much later. But we know almost nothing about their lives, and it is clear that not one of them took an interest in history, despite the proximity of civilizations that did.

In the sixth century we also know something, though hardly much, about Solon, the legislator, and Pisistratos, the benevolent tyrant; but Greek history more or less begins with the murder of Pisistratos' son, Hipparchos, and the expulsion from Athens of his other son, Hippias, in 510. Oddly, the Athenian democracy was born, or at least restored, the very same year that the Romans ended the rule of the Etruscan kings and established the Roman republic.

What individuals do we know of before 600 B.C.? Perhaps none outside the Near East. In China and India, Iran and Europe, America and Africa, with the exception of Egypt, none. But in Egypt, Mesopotamia, and Israel, many; and more in Israel, despite its small size, than in Egypt and Mesopotamia taken together.

Not that there were no remarkable cultures elsewhere. In China, magnificent bronzes were made before 1,000 B.C.; in Mesoamerica, superb sculptures; and some of the cave paintings in southern France and northern Spain are breathtaking. One could lengthen this list and add, for example, the poems of Hesiod in Greece; but one would still be at a loss to add the names of men and women who were contemporary with these creations and whose lives and characters are known to us.

The closest that we can come to an apparent exception is Homer. Actually, we know nothing of his life or character, and most scholars agree that the two poems ascribed to him reflect widely different outlooks and could not really have been composed by the same man. Even the notion that he wrote the *Iliad* when young and the *Odyssey* in his old age is clearly untenable. But that still leaves the question whether these epics do not portray individuals: notably Achilles and Odysseus, but also Agamemnon and Ajax, Hector and Helen, Priam and Penelope, and perhaps quite a number of others. Indeed, some of us grew up with these characters, and they have formed part of the world of our imagination for almost as long as we can remember.

Friedrich Schiller's poems on some Homeric themes, like "The Victory Festival" (*Das Sieges-fest*) and "Hector's Farewell" (*Hektors Abschied*), were symptoms of and contributed to "The Tyranny of Greece over Germany," to cite the title of a fine book by E. M. Butler. But what Schiller did in a small way the tragic poets of Greece had done on a large scale. They prepared their poetic feasts from what Aeschylus himself called "slices from the great banquets of Homer."*

Today scarcely anyone except a few specialists knows any more what information about Achilles, Agamemnon, or Ajax is to be found in Homer, and what is first found in fifth century tragedies or in still later writers. And none of these poets were concerned to find out what these men and women of the remote past had really been like. Their interest was not at all in history. Sophocles, one of the greatest poets of all time, had no compunctions at all about giving Odysseus one sort of character in his *Ajax* and quite another in his *Philoctetes*. He was guided solely by his dramatic requirements.

Of Sophocles' one hundred and twenty plays, only seven survive; of Euripides', at least nineteen. Hence we find far more "information" in the plays of Euripides, and also many similar examples. These poets were not concerned with history or what really happened. As Aristotle said, "The historian relates what happened, the poet what might happen. . . . Poetry deals with general truths, history with specific events."

It is a commonplace that Heinrich Schliemann, the German amateur archaeologist, dug up Troy and Mycene and proved that the *Iliad* and the *Odyssey* contain accurate memories of the distant past. No doubt about it, but one can approach such memories as a historian would—trying to check them in an attempt to discover what really happened—or one can use them as a poet might—the way Sophocles did, and Homer as well.

11

The Greeks had surprisingly little interest in history. Thucydides did have the ethos of checking reports and told his readers at the beginning of his *History of the Peloponnesian War* that he had relied "neither on arbitrary hear-say nor on my own suppositions, but on my own experience or on reports that I have checked carefully in every detail."

Proud of his own critical spirit, Thucydides voiced his contempt for those who lacked his ethos: "So averse to taking pains are most men in the search for truth, and so prone are they to accept what lies ready at hand" (I.20). In many ways Thucydides remains exemplary; but in a sense even he did not write real history, for he recorded what happened in his own lifetime and did not reconstruct the past on the basis of documents. Nor did he have a larger, more comprehensive conception of history in which the war he recorded was merely one chapter. Even he did not think in terms of long-range developments, nor did he aim to explain fundamental changes. He says at the outset, speaking of himself in the third person: "Thucydides, the Athenian, records in this work the war between Peloponnesians and the Athenians. He began his description immediately when the war started because he foresaw that it would become a great war, the most memorable of all that ever occurred."

In fact, he knew nothing of most wars of the past, and in an important sense he lacked historical perspective. Even the story of a single great war can be told with an eye to character development, focusing attention on the ways in which some representative people changed. The point of the title of *Gone with the Wind* is that after the Civil War a whole way of life had disappeared, and that the people who survived the war were changed profoundly. It is arguable that it is the historian's task to explain in some detail how changes of this sort were brought about, and that this is done best when he operates with a larger framework, seeing a war, for example, as one stage in the development of a people or a culture.

It would go much too far to say that Thucydides' philosophy was Eleatic-Platonic. But considering that he was by far the greatest historian the Greeks produced, it remains remarkable how little interest he showed in change and time. Like Herodotus and the poets of Greece, he was a great story teller, but it is by no means incomprehensible why Aristotle, who knew Greek tragedy, said, "Poetry is more philosophical and nobler than history."

* Athenaeus, *The Deipnosophists*, VIII. 347 E.

A twentieth-century British classicist devoted a whole book to showing that Thucydides "did not, as is commonly asserted, take a scientific view of human history" (F. M. Cornford, *Thucydides Mythistoricus*, p. ix). Specifically, Cornford aimed to show that "Thucydides' Conception of History" (that is the title of Chapter V) did not encompass the *causes* of events, but only events and speeches. Regarding the speeches, Thucydides himself said: "It would be hard to reproduce the exact language used, whether I heard it myself or it was reported to me by others. The speeches as they stand represent what, in my opinion, was most necessary to be said by the several speakers about the matter in question *at the moment* [my italics], and I have kept as closely as possible to the general sense of what was really said." Cornford has emphasized that this is really "an extraordinary method," and that Thucydides, instead of discussing causes, has "put us off with the *ex parte* [that is, partisan] 'accounts' of interested persons, as publicly and formally stated with a view to persuading other interested persons" (p. 54). As soon as one considers the *History of the Peloponnesian War* in this perspective, it becomes evident that Thucydides took his cue from the great tragic poets—without, of course, relying *entirely* on speeches. Cornford further tried to show how much of the work depends on a view of human nature that "Thucydides seems to me to have learnt from Aeschylus" (p. x). But nothing in the present book depends on Cornford's theses. They are mentioned only because, from a different vantage point, they call into question some widely shared but ill-founded prejudices.

12

In the *Iliad* and the *Odyssey* there is no character development. The heroes of these epics, Achilles and Odysseus, are magnificent and unforgettable, but they are types, not individuals. Odysseus, about whom we are given far more information, can actually be defined quite easily as a man with three qualities which he possesses in the highest degree: toughness, courage, and cunning. There is no end to the adventures one can attribute to him, but none of them change him. He is timeless.

Achilles is defined as the hero unequaled in prowess and wrath. In his anger he even forgets his chivalry. The folktale that the great hero was invulnerable save for one spot—in his case the heel—was not connected with him before the first century A.D. and is first found in the *Achilleis* of Publius Papinius Statius, a Roman poet. This tale detracts from Achilles' valor and is quite out of keeping with the spirit of the *Iliad*, in which he himself says that the price he must pay for immortal glory is an early death, though he could return home instead of fighting, and live to old age without any glory (IX.410ff.). That is part of his prowess and helps to make him the most glorious of all the great heroes.

The *Iliad* and the *Odyssey* left their imprint upon the Athenian imagination, by no means only upon the tragic poets. How can we explain the fact that after centuries of dark ages, in which the Greeks produced marvelous pottery, and toward the end also some splendid sculptures, but nowhere near half a dozen remarkable individuals—perhaps only two: Solon and Pisistratus—one wearies of counting all the memorable human beings of fifth-century Athens? Something must have happened in the later sixth century to produce this immense explosion.

Like the Greeks themselves, most people are so unhistorically minded that the very question has never occurred to them. But once we ask it, an answer suggests itself. Before he died around 527, Pisistratus had the two great Homeric epics assembled and introduced them at Athens in the form in which we still know them. I surmise that he loved them because he found them profoundly congenial, and his own astonishing character may have been formed by them. It is well known that in the last half of the fourth century Alexander's ambition was fired by Achilles, who had won immortal glory so young; and it is more than likely that Aeschylus, Sophocles, and Euripides were not the only ones whose ideas, feelings, and works are unthinkable without Homer. That the sudden explosion of first-rate poetry was due to the impact of Homer is obvious. What I am suggesting is that the effect was not confined to the tragic poets but felt throughout Athens. Of course, it was a snowball effect to which Aeschylus contributed his share, and then, inspired by men vied first with Pisistratus and then also with Themistocles, Pericles, and so forth.
him as well as by Homer, Sophocles; and at the same time other statesmen and would-be states-

In this great cast of real-life characters none stands out as more original and unforgettable than Socrates. He did not model himself on any hero in Homer or any previous philosopher;

he created a new type; he invented himself. But he, too, is a type like Achilles and Odysseus, a timeless type who had a few, really very few, striking experiences. By far the most impressive of these is how he was sentenced to death and died; but no Greek asked how the man might have become the way he was.

13

When we look at Greek tragedy in this perspective, we find that, even here change is curiously circumscribed. The characters do not develop. Sophocles' *Antigone*, perhaps as great a play as has ever been written, shows this very clearly. The two central characters, Creon and Antigone, embody different outlooks, and the play deals with their tragic clash. Neither of them changes. Antigone, the quintessence of noble defiance, hangs herself; Creon, confronted with her death as well as his son's and his wife's, breaks down.

In *Oedipus Tyrannus*, which is also second to none, the protagonist blinds himself in the end as his wrath flares up against himself. He has learned who he is, that his wife is his mother, and that he has killed his father, but all of this does not change his character.

These plays have a ritual quality. Greek tragedy reenacts, every time it is performed, some gruesome events that are repeatable again and again like harvest festivals. The effect is in some sense cathartic, as Aristotle noted in his *Poetics*. These tragedies are immensely noble and philosophical, and as he also noted, anti-historical in their ethos. They dramatize timeless conflicts or show us tragic conflicts as timeless.

It should come as no surprise that fifth-century Athens was based on slavery, and that Aristotle, who taught Alexander, still thought of the distinction between slaves and free men as eternal no less than the fixety of species. The Greeks, he thought, were by nature free; the barbarians, meaning all non-Greeks, slaves. But Alexander did not accept this doctrine.

14

Perhaps none of the great Athenians came closer to an appreciation of time and a conception of history than Aeschylus. Like his successors, he wrote trilogies; but theirs consisted of three independent tragedies, while his often told a single story in three plays. Only one such trilogy has survived intact: *The Oresteia*. It has long been recognized as one of the greatest achievements of world literature, but it also approximates a historical vision.

In *Agamemnon*, the first of the three plays, the chorus says twice in its first long speech or song that one learns through suffering. In the Old Testament such an aphorism might scarcely be noted because this theme pervades the whole of it. In Athens, however, it is like a flash of lightning in a cloudless sky. One is tempted to call it un-Greek. The remark is made within hearing of Clytemnestra; yet her suffering teaches her nothing, and the plot of the play depends on the fact that Agamemnon, too, has acquired no wisdom through all his tribulations. None of the impressive principal characters in this trilogy becomes wise through suffering, and one is led to wonder whether the nub of Greek tragedy is not precisely the failure of men and women to learn through suffering. Aeschylus in particular seems to say again and again that the catastrophes he describes were not at all inevitable. They could have been prevented if only people had learned from their suffering.

The first great dramatist did what Bert Brecht, after World War I, claimed as his own innovation. Aeschylus confronted his audience as a teacher. Though his tragedies had a cathartic effect—and Brecht's *Mother Courage* has that, too—he clearly aimed also to be an educator, and his example was followed by many great dramatists before Brecht, including Euripides, Lessing, Schiller, Ibsen, and Shaw.

Poetically, modern readers are almost bound to find the first play of *The Oresteia* by far the most moving, and the last play downright anticlimactic. But it is the trilogy as a whole that shows us the poet's proto-historical vision. From play to play, the climate of thought and feeling changes, as we move from level to level of a grand development that issues in the founding of the Areopagus, the Supreme Court of Athens. Of course, the origin of this institution was nothing like Aeschylus' account of it. Still, instead of assuming that the court had always existed, he viewed it as the result of human experiences and sought to show why it was needed.

If any characters change in *The Oresteia*, it is the Furies. These inhuman, superhuman spirits of vengeance, who, with snakes in their hair, hound Orestes, the matricide, cease being terror incarnate and become the Eumenides. They are the chorus of the last play, which is named after them. They change as vengeance is sublimated into the justice administered by a human court.

This took time and was, in a sense, the work of time. In the *Prometheus* trilogy, of which only the first part survives intact, the plot seems to have followed similar lines. It took time for Zeus and Prometheus to learn that, in their very own interest, they must come to terms. Their characters did not change; but that they came to see reason was still the work of time.

In a study of the discovery of the individual, Euripides would require a great deal of attention. He explored, like no one before him, the psychological dimension and the abundant irrationality of men and women. But he, too, did not study how human characters change. Like Aeschylus, he bade us weep that they do not change.

15

Heraclitus, Aeschylus' older contemporary, may have been the first man to reflect expressly on the nature of time. He lived in Ephesus in what now is Turkey but was then, around 500 B.C., part of the Persian empire. No other early Greek philosopher had his gift for composing haunting aphorisms. "Character is man's fate."

He derided the views of others, said that "men are like people of no experience," expressed scorn for Hesiod and Pythagoras, and wrote in another fragment that has survived from his book that "Homer deserves to be thrown out of the contests and whipped." He is remembered above all for insisting that all is in flux, and for saying, "It is not possible to step twice into the same river."

He also made much of fire, and one fragment reads: "This cosmos none of gods or men made; but it always was and is and shall be: an everlasting fire, kindling in measures and going out in measures." He probably held an ultimately cyclical view, as later philosophers thought he did, and his fire fragments seem to be influenced by Zoroastrianism and its vision of a fiery last judgment.

"Time is a child playing a game of draughts; the kingship is in the hands of a child."* We cannot be sure what Heraclitus meant by this. The ancients called him "the dark philosopher" and found much he said very obscure. We have only scattered quotations in ancient writers and a few remarks about him, and scholars do not agree about his ideas. He may have meant that time is not rational, that the world is not governed by any purpose, and that man's lot is absurd.

It may be objected that this is a modern misreading which turns a Greek who lived 2,500 years ago into an existentialist. To this, one might well reply that after Platonism and Christianity lost their credit, some modern thinkers recovered the wisdom of Heraclitus.

Heraclitus did not present "his arguments only to be refuted," as Leo Baeck claimed. In his surviving fragments we find no arguments, nor did he sing the praises of time and change. Like the philosophers of Elea, he rejected not only Homer and Hesiod and all of the philosophers who came before him but also common sense. His thinking was not historical.

16

The Jews changed human attitudes more than any other ancient people. They experienced time differently. They overcame the widespread, if not universal, terror of change.

To adapt to change and survive, one must note, not necessarily consciously, some regularities, some recurrent patterns. Even animals do that. And human beings find change less frightening when it is regular, and if possible, cyclical. Sunrise and sunset, dawn and dusk, day giving way to night, or night to day, are not fraught with horror if one knows before *that* this will happen, and *when*, and that night will give way to day again, and day to night. There is then, after all, no *real* change, since the whole cycle is always the same. It is not unpredictable and does not seem to be irrevocable, like fire. But the passage of time *is* irrevocable, and most men have always been extremely reluctant to face up to this truth.

Heraclitus made much of change and even of fire, but even he was no exception, and fragment 30, already cited, reads like a rejoinder to the Old Testament and a reiteration of the old pre-Hebrew wisdom: "This cosmos none of gods or men made; but it always was and is and

* The quotations are from fragments #119, 1, 42, 91, 30, and 52.

shall be: an everlasting fire, kindling in measures and going out in measures." In this view, even fire is not irrevocable. All is regular, all recurs.

The Jews introduced a radically different view. One need not look far and wide to glean it from obscure texts. It animates the Hebrew Bible from the first word to the last. One might think that nobody could miss it, but not seeing the wood for the trees in it has become the academic epidemic, and scholars who write on the Bible try to earn their spurs by focusing attention on details and arguing, for example, that this verse or half verse was written later than that. Formerly, when knowledge of the Bible was much more widespread, the scholastics' endemic disregard for context dominated all references to the Bible. Verses and even phrases were ripped from here and there and used like so many twigs and leaves to build one's own nest. Moreover, one had no conception in those days of radically different cultures, and hence could not possibly realize how the Jews had introduced radically new ideas. Though Plato and Aristotle were not unknown, Christians all but treated them as church fathers.

One might suppose that the first words of the Bible must have struck people like the opening chords of Beethoven's Fifth Symphony, but they did not. "In the beginning God created the heaven and the earth."* There was a beginning! And from there the story proceeds relentlessly, relating what happened just once. Here, surely, is the origin of this passage in Rilke's Ninth Duino Elegy:

> . . . *Once*
> everything, only *once. Once* and no more. And we, too,
> *once.* Never again. But having
> been this *once*, even though only *once:*
> having been on earth does not seem revokable.

17

The theme announced in the first words of Genesis is the ethos of unique events. Things have not always been as they are and shall be. There is genuine change. Festivals, institutions, peoples, and even the earth and the heavens have a beginning, and men and women make irrevocable choices that change their lives as well as the lives of their descendants.

It may be objected that what we find in Genesis is not the oldest creation myth, and this is true. But look at the ones that are older. *Ancient Near Eastern Texts Relating to the Old Testament,* edited by James Pritchard, makes it easy to do that and shows how different Genesis is from all books that preceded it. *Ancient Near Eastern Texts* begins with Egyptian myths, and the first, which the editor of this section calls "The Creation of Atum" is a text that "served in the dedication ritual of a royal pyramid by recalling the first creation." It begins:

> Atum-Kheprer, thou wast on high on the (primeval) hill; thou didst arise as the *ben*-bird of the *ben*-stone in the *Ben*-House in Heliopolis; thou didst spit out what was Shu, thou didst sputter out what was Tefnut. Thou didst put thy arms about them as the arms of a *ka*, for thy *ka* was in them. (So also), O Atum, put thou thy arms about King Nefer-ka-Re, about this construction work, about this pyramid . . .

There is surely no need to quote the remaining two thirds. And the next text, which the editor calls "Another Version of the Creation of Atum," also does not steal the thunder of Genesis.

We come a little closer to Genesis in the Creation Epic that opens the collection of "Akkadian Myths and Epics." This is a substantial text. Not quite all of it has survived, but the translation comprises more than 900 lines: It begins:

> When on high the heaven had not been named,
> Firm ground below had not been called by name,
> Naught but primordial Apsu, their begetter,
> (And) Mummu-Tiamat, she who bore them all,
> Their waters commingling as a single body;
> No reed hut had been matted . . .

The effect is numbing and becomes more and more so. The comparison increases one's appreciation of the sublime economy of Genesis and of its radically different ethos. In the texts of the ancient Near East, of India, and of Greece one looks in vain for any narrative that is half as terse and succinct as Genesis, or that approximates its sustained narrative sweep.

---

* In some recent versions of Genesis the first verse becomes the opening clause of a long and involved sentence, although all that follows is laconic.

18

The first ten and a half chapters of the Hebrew Bible are amazingly different from anything written earlier or, for that matter, later. These ten pages have enriched man's imagination with the stories of the creation, the so-called fall, the expulsion from paradise, Cain and Abel, the flood, and the tower of Babel.

After that the pace slackens. Still in chapter 11, Abraham is introduced, and all the rest of Genesis deals with him and his family, down to his great-grandson Joseph. Indeed, the second half of Genesis is dominated by a single human being, Jacob. But the change of tempo involves no decrease in originality. The story of Jacob is the first example of character development in world literature.

Even if one assumes that Genesis was written as late as 500 B.C. it would still amply support this claim; and since the dating of Genesis is steeped in controversy, one might let it go at that. But it may be interesting to note that the Jacob stories must have received something like their present form several centuries before the destruction of the kingdom of Israel in 722 B.C. If the central emphasis on Joseph as Jacob's favorite son had not been hallowed by a very old tradition, it would certainly have been changed when the ten tribes, including Joseph's descendants, disappeared. Judah, the eponymous hero of the southern kingdom which survived the disappearance of the ten tribes, would have supplanted Joseph as the favorite.

Moreover, the Jacob stories must antedate the division of the united monarchy in 933 B.C. For they obviously cannot have originated in Judah, and if they had been composed in the northern kingdom, which seceded in 933, they would surely contain some reflection of the rivalry between the brothers after whom the northern tribes were named and those who founded the southern tribes, Judah and Benjamin.

Finally, the preference for the younger brother (Isaac over Ishmael; Jacob over Esau; Joseph over his brothers; and eventually, in Jacob's blessing of Joseph's son, Ephraim over Manasseh) is one of the leitmotifs of Genesis and would have made it natural to hint that Saul, the first king of the united monarchy, was of the tribe of Benjamin. The absence of any suggestion whatever that Benjamin's descendants would ever do anything special makes it probable that the stories antedate the eleventh century, and received something like their present form during the reign of David and Solomon, around the year 1000, when the scepter had only just passed to Judah. Many Bible scholars still think in terms of a much later fusion of at least two ancient sources by an editor who felt free to take great liberties. The arguments offered here show why this is highly unlikely. But for our purposes here, a few centuries one way or the other scarcely matter.

Jacob is not a hero from the start. One can say of Achilles in the *Iliad*—which is far longer than the second half of Genesis—and of Odysseus in both of the great epic poems, that they are defined by such and such traits—actually, not more than three or four in each case. Every time Joseph K. opens a certain door in Kafka's *Trial*, there is a man who is being whipped. One might say that every time one looks at a Homeric hero one can count on his still having the same characteristics. All his experiences do not change him. It is the same in Greek tragedy. But Jacob is different. We see him as a mother's boy, very unlike his brother, Esau, who is a hunter. Jacob is somewhat fearful, crafty, and no fighter. But in a terrifying encounter he rises to the occasion and becomes a hero.

> When dark had sealed his solitude, and dawn
> was likely to bring death, he did not say,
> Remove this cup, but let thy will be done!
> He wrestled God until the sky grew grey,
>
> and when God saw that He could not prevail
> and wounded him and pleaded, Let me go!
> a weary man that once was small and frail
> mastered his terror and forgot his awe
>
> and, seeing God, said, I will not unless—
> my father said, my blood, my flesh and bone,
> Jacob no more but Israel—you bless
> me who has fought and conquered God, alone.

This contemporary retelling of the tale* is closer to Genesis 32 than the popular notion that Jacob wrestled with an angel. There is no angel in the original: "Jacob called the name of the

---

* "Jacob" in Kaufmann, *Cain and Other Poems*.

place Peniel, saying, 'For I have seen God face to face' "; moreover, his name is changed to Israel because "you have fought God and men and have prevailed." The new name, Israel, is taken to mean God-fighter.

After that, Jacob is no longer a man who tries to make up for weakness by craft. But he does not simply change from one identity to another; he grows old. The book shows us a man growing old and displays an extraordinary sensitivity for the feelings of a father and then of an aged man.

19

The Greeks proved that literature can be first-rate without offering character development. Scores of contemporary examples show that literature can be wretched while offering it. But the first writer to show us how time is an artist was a supreme artist.

The young Jacob was a mover who did things. As he ages, he suffers. His beloved Rachel, the wife for whom he served twice seven years, dies. Her son Joseph, his favorite child, is sold into slavery by his envious brothers who lead the old father to believe that he has been killed. "And he tore his garments, put sackcloth on his loins, and mourned his son many days."

Joseph becomes the most powerful man in Egypt next to the pharaoh. His brothers leave home in a famine and come to ask Joseph for grain, but do not recognize him. He demands that next time they bring along Benjamin, Rachel's other son and Jacob's youngest. While the others return home, one of the brothers, Simeon, has to stay behind in Egypt. The old Jacob does not want to let Benjamin go: "My son shall not go down with you, for his brother is dead, and he alone is left. Should he come to grief on the journey you will make, you would bring down my grey head with sorrow to Sheol." But the famine worsens, and there is no way out but for the sons to go to Egypt again, with Benjamin. Jacob must let him go and says: "May Almighty God [El Shadday] grant you mercy before the man to send back with you your other brother and Benjamin. And I, if I am bereaved I am bereaved."

This time Joseph's first question to them is: "How is your father, the old man of whom you spoke? Is he still living?" The whole story is told so simply and movingly that it remains unsurpassed.* Eventually, Judah, one of the brothers, explains to Joseph, still unaware who he is, how Jacob feels about Benjamin and about Joseph. And Joseph, who is unable to hold back his feelings any longer, sends out everyone except his brothers and reveals himself to them. Then he charges them: "Hurry and go up to my father and say to him: Thus speaks your son Joseph: God has made me master of all of Egypt, come down to me, do not delay." But when they get back and tell their old father, "his heart stayed numb, for he did not believe them. Then they told him all the words of Joseph that he had spoken to them, and he saw the wagons Joseph had sent to carry him. Then the spirit of Jacob, their father, revived, and Israel [note the change of the name at this point: he becomes a fighter again] said: Enough! My son Joseph is still alive. I want to go and see him before I die."

The evocation of old age is as moving as the narrator's feeling for the father's love. Never mind the boy's great position and all the provisions he has promised to make for his father and brothers in Egypt. "Enough! My son Joseph is still alive. I want to go and see him before I die."

The new ethos introduced in the opening words of the book, and developed through the story of man's creation in the image of God, becomes more and more concrete in the following chapters. The distinctive conception of man and his dignity and of time and change does not remain at the symbolic level; it gradually pervades the reader's feelings and creates a new sensibility.

Men do things, just once, that they might later wish they had not done; but their acts are irrevocable and have unforeseen consequences. Jacob's aging is irrevocable also, but the process is not merely one of decline and deterioration. He grows in dignity, stature, and significance as he becomes a patriarch. Even Joseph's rise to incredible power is but a chapter in Jacob's life. God is the God of Abraham, Isaac, and Jacob, and the Jews are seen as the children of Israel.

20

The ties of Israel to the ancient civilizations of Egypt and Mesopotamia are stressed in Genesis; and Exodus, which continues the story, make much of Moses' access to the learning

---

* Eric Lowenthal's book should convince anyone of that.

of Egypt. But the Egyptians, like most men, refused to see change as irrevocable and denied death's finality. Their immensely impressive culture was dominated by their concern for the afterlife. And they, like other ancient peoples, did not see all human beings as brothers and sisters made in the image of one God and descended from one human couple. The pharaoh was a living god; the priests were learned and literate, most of the people were not; and the state was founded on slavery. The Hebrew idea that one should love the stranger (Leviticus 19) was as foreign to the Egyptians as it was to other ancient civilizations.

For the Egyptians, who built the pyramids and made mummies, time was the great enemy. The Hebrews accepted death as final, turned their back on the visual arts, and made peace with time. Change was recognized as the way of the world and accepted. Man must live with it.

Time was not viewed as a child playing a game of draughts, but as the dimension of history. And history was a story that began with God's creation of heaven and earth and the other events related in Genesis, and it continued with the exodus from Egypt and God's revelation at Mount Sinai. It included the death of Moses. Deuteronomy noted expressly that no man knew his grave, but it was not doubted that he was dead and buried.

History also included the conquest of the promised land, the period of the Judges, including Deborah, Samson, and Samuel, and the stories of the kings: Saul, David, Solomon, and their successors. History was seen as a development, as a succession of great changes, and as the context in which real men and real women lived, suffered, grew old, and died, often leaving behind a striking legacy.

21

Many men are seen to grow in stature as they age. Samson at first seems to be a type, a folk hero defined by a few traits, like Achilles and Odysseus. He is enormously strong, a great fighter who repeatedly saves his people, but the Philistine women he beds worm his secrets out of him with disastrous results. Such a story might be found elsewhere, though the narrative art with its sublime economy and its consistent refusal to engage in epic digressions remains distinctive. But as the story progresses relentlessly, the unrepeatable is introduced: the hero is blinded, subjected to suffering, and in his death grows to unforgettable stature.

Samson is the first of the four suicides in the Hebrew Bible. He breaks the columns that support the roof under which he is mocked and tormented, and "the dead he slew dying exceeded those he had slain living." Reduced to slavery and blinded, he refuses to give up and grow old like an ox. He can still make choices and act, and inflict an irrevocable death on his enemies and himself.

A contrast with Oedipus is revealing. Rather than having a single Oedipus story, we have several different versions. The oldest, in the eleventh canto of the *Odyssey* (271–80), is quite different from those of Sophocles.* Sophocles dealt with the theme in two of his tragedies: *Oedipus Tyrannus*, around 425, and *Oedipus at Colonus*, finished in 406, just before the poet died at the age of ninety. Modern translators have sometimes presented these plays as parts of a single trilogy, with *Antigone*, which was first performed around 442, thrown in as the final play. But a poet who wrote 120 plays had to deal with the major myths more than once, and any notion that the tragedies which revolve around some of the same characters should be viewed as more or less parts of a single work is untenable. As mentioned earlier, Odysseus appears in two of Sophocles' tragedies but is not at all the same character in the two, and nobody would even attempt to seek any character development here. From Euripides' hand more tragedies have survived than from Sophocles'. Orestes appears in several, and again it is plain that we cannot arrange them in any sequence or trace a development here.

In neither of Sophocles' Oedipus tragedies does the character of the hero develop. In each he remains the same throughout. At the end of the first, be blinds himself as his uncontrollable anger is directed toward himself. In the late play he is blind and, as it happens, still given to outbursts of wrath. In the end he is somehow transfigured, but not by virtue of anything he does. The lines between men and gods were not drawn sharply in ancient Greece. At the end of Sophocles' *The Women of Trachis*, Heracles, who also had not grown through his terrible suffering, is elevated among the gods. The ethos is different from that of the Bible: death is not irrevocable. Samson's death will be final, and he knows it.

* For comparisons of all the major Greek versions, see Kaufmann, *Tragedy and Philosophy*, section 22.

22

The last of the judges, Samuel, anointed Saul to become the first of the kings; and later, rejecting Saul, he anointed David to be his successor. In length the story of Samuel, Saul, and David exceeds the Book of Genesis. It embodies a good deal of genuine history, and Eduard Meyer (1855–1930) considered this part of the Bible the first genuine example of historiography anywhere.

Meyer wrote a multi-volume history of antiquity (*Geschichte des Altertums*), devoting roughly two volumes each to Egypt, Mesopotamia, and the Persians and Greeks; and he also published many books of Bible criticism. His mastery of the whole field is as striking as his pronounced distaste for Judaism. Despite this antipathy, Meyer found that the Jews were the first to write genuine history. He distinguished royal inscriptions and even consecutive annals from true historiography. Section 131 of the first volume of his *Geschichte* begins:

> All this is not historical literature in the true sense of that word. Of that one cannot even speak when, as happened among the Egyptians and Babylonians, single historical legends are written down and given literary treatment, or when historical documents are collected and copied, as happened in the library of Assurbanipal, or when, as happened among all of these people, consecutive chronicles are maintained and a general overview of the traditional history of the people becomes a common property of at least the more highly educated, meaning the rulers and priests. It really comes into being only when . . . single individuals make it their life's vocation to collect and work over the traditions to make of them an autonomous and unified historical work that bears the stamp of their individuality. This historical sense has developed autonomously only among a very few people. Even people who had a very high culture and eminent historical importance, like the Aryans of Iran and India, have failed to produce not only a historical literature . . . but even chronicles after the manner of the Egyptians and Babylonians, because their conception of the existence and evolution of man was overgrown entirely by myths (with components of legends) and religious notions. . . . A completely autonomous creation of a truly historical literature occurred in the area of the Near Eastern-European culture only among the Israelites and the Greeks. Among the Israelites, who in this respect, too, occupy a unique position among all the civilized peoples of the Orient, it came into being at an amazingly early time and began with creations of the highest rank, first the purely historical narratives of Judges and Samuel, and then the working over of legends by the Yahwist [the putative author of those portions of the Five Books of Moses in which *Yahweh* appears as the name of God; the postulate that there was a Yahwist was a defining characteristic of the so-called Higher Bible Criticism]. But its further development was stunted as the religious development that culminated in Judaism smothered it. Among the Greeks it originated only at a much later stage . . .

It is striking how the stories of Samuel, Saul, and David, who are clearly historical characters, are informed by the same ethos as Genesis. These men are truly descendants of Jacob, men whom we get to know before they are heroes, whom we see as they grow, participate in momentous events, age, suffer, and die. Each of them is an individual, to be known, not through a short string of attributes, but through a long life in the course of which he develops.

The old Samuel, Saul, and David are not merely the same men we have known in their youth after a few more adventures; they really are *old* men who show the effects of time and grief, men cruelly shaped by time's artistry.

23

For the Jewish people, the children of Israel, there is no death, and even the Babylonian exile is not the end. The Hebrew Bible ends with the words of Cyrus of Persia, who conquered Babylon and allowed the Jews to return to Jerusalem: "Whoever is among you of all his people, may the Lord his God be with him. Let him go up."

The Jews returned, but there was no "restoration." The temple was rebuilt eventually, but it was not like Solomon's temple. There was no going back to the *status quo ante*, to a past that was gone irrevocably.

The children of Israel after the exile were not like the children of Israel before it, any more than the people of David's time were like Jacob and his family when they moved to Egypt. Time made a difference, change was real, and the point was not to decide whether things had once been far better or whether perhaps they used to be worse. The present was always a chapter in a long story that began with the Book of Genesis. And though one did not know what might happen a hundred years hence, one did not suppose that the present was the conclusion. The story would continue.

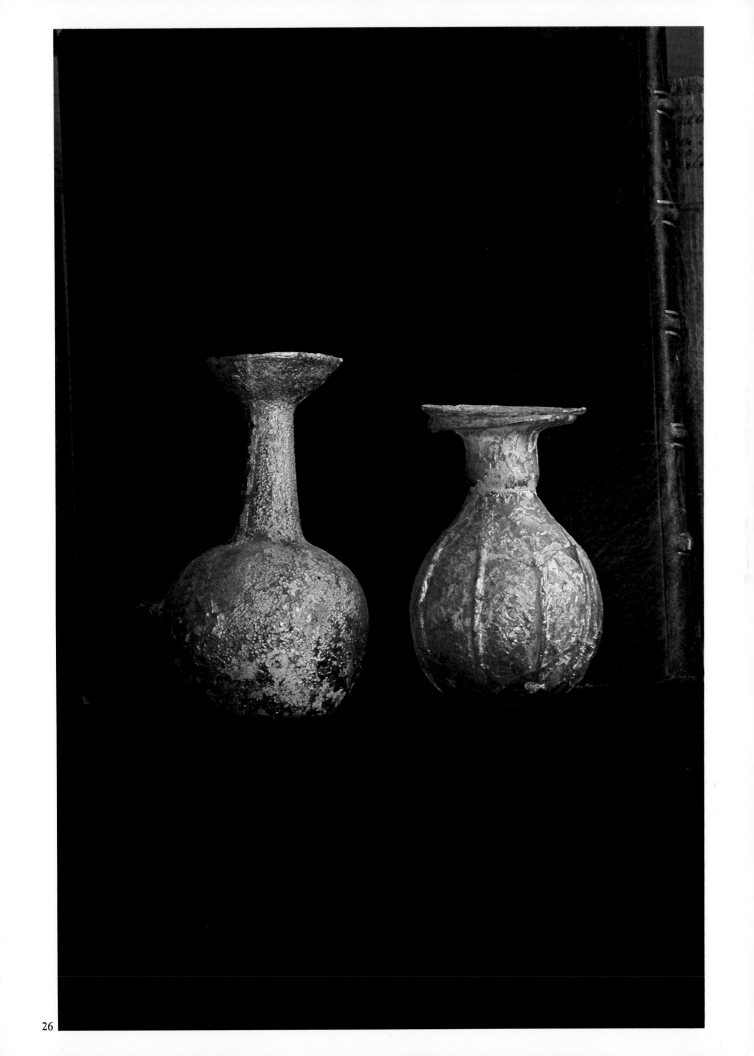

29

30

47

48

49

No other ancient people saw itself this way. Even the Greeks did not have this keen historical consciousness. And at the heart of the Jewish vision of history was the Jews' sense that they had a task they had not completed, a task that required time. To understand this mission one has to reread Genesis and study Moses, the judges, the kings, and the prophets.

The reaction of Jeremiah and the Second Isaiah (the author of Isaiah 40ff.) to the destruction of Jerusalem was remarkably different from Plato's to the defeat of Athens in the Peloponnesian War. Far from feeling that Athens still had a mission, Plato concluded that its democratic form of government must have been vastly inferior to the government of the victorious Spartans.

24

In modern times, one might think, no great people is more unhistorical than the Americans. They are newcomers on the stage of history. And it is true that most Americans do not have much historical sense. Their conception of what is "old" sometimes borders on the ridiculous, and anything over a hundred years old is considered an "antique."

Yet it was not only the prose of their greatest and most beloved statesman that was nourished on the Old Testament but his conception of history also. In a period of immense suffering he was sustained by the faith that his people was "the last, best hope on earth," that it had a mission and dare not fail, that there was "unfinished work" and a "great task remaining before us." What saved such proud words from sounding insufferably arrogant was his keen sense of suffering and his overwhelming feeling that he must do his share to make sure "that these dead shall not have died in vain."

It is a measure of Abraham Lincoln's depth that other wartime Presidents and world leaders seem shallow compared to him. None of them was so profoundly affected by the suffering entailed by war, and some actually seemed to relish the roles they were allowed to play in vast international crises. Lincoln not only found words for the anguish he shared with the maimed, the widows, and the orphans, but his own face changed in the course of the war. As we look at photographs of him in chronological order, we see that time is an artist.

25

Why is it that we do not know how any of the great thinkers of ancient Israel, India, Greece, and China looked? We have portraits of Sesostris III, an Egyptian pharaoh of the eighteenth century B.C., and have no difficulty recognizing his features in a museum when we see a sculpture of him that we had not known before. Of Akhenaton, the heretical pharaoh of the fourteenth century, we have more different portraits than of any other man before Rembrandt. Yet we have no contemporary portraits of the Buddha, Confucius, Lao-tze, the tragic poets of Greece, Socrates, or the ancient Hebrews whose deeds and words we know from the Bible. How can we explain this?

The reasons are different in each of the four cases. In India, the lack of interest in the ephemeral led to the total absence of portrait art until modern times, when Mogul (Mughal) painters who did meticulous miniatures began to paint portraits in the sixteenth century of our era. In his *History of Indian and Indonesian Art*, Ananda Coomaraswamy said:

> It is important to understand the relation of Rajput to Mughal painting. Pure types of either can be distinguished at a glance, usually by their themes, always by their style. Thus Mughal painting, like the contemporary Memoirs of the great Mughals, reflects an interest that is exclusively in persons and events; it is essentially an art of portraiture and chronicle. The attitude even of the painters to their work is personal; the names of at least a hundred Mughal painters are known from their signatures, while of Rajput painters it would be hard to mention the names of half a dozen, and I know of only two signed and dated examples.

Coomaraswamy, who was curator of Indian and Muslim art in the Boston Museum of Fine Arts, goes on to quote Jahangir, the Muslim emperor whose son Shah Jehan built the Taj Mahal. When Jahangir was still a prince, he expressed his feelings about Hindu art in these words: "The old songs weary my heart . . . if we read at all, let it be what we have seen and beheld ourselves." And Coomaraswamy adds that Rajput painting 'illustrates every phase of mediaeval Hindi literature, and indeed, its themes cannot be understood without a thorough knowledge of the Indian epics . . ." (p. 127f.)

Oddly, the emergence of portrait painting in India can thus be traced to the Old Testament! By way of Islam, the interest "in persons and events" eventually came to northern India. This is ironical, because in the Hebrew Bible and in early Islam we encounter strict prohibitions against making human likenesses. But the reason for this was, of course, to prohibit idolatry, and the miniature painting cultivated by Muslim artists under Persian influence did not facilitate or promote that.

Thus the reasons for the absence of portraits and the neglect of the visual arts in ancient India and Israel are almost diametrical opposites. In India, the Brahmins opposed all interest in the particular, and we find no interest in or recognition of individuals. Eventually, sculpture and painting proliferated, but the old frescoes, reliefs, and sculptures do not bring us face to face with unique individuals or events, or with what the artists had seen and beheld themselves. The subjects are always timeless and in a profound sense escapist. What is portrayed is neither the beauty nor the misery of this world, but gods and scenes from old epics. Khushwant Singh's critique of Indian films, "We sell them dreams," applies to almost all of Indian art.

26

In China there was some recognition of individuals, but on reflection, surprisingly little. Though in some ways the Chinese had a very high civilization from the late second millennium onward, Confucius is the first historical human being of whose outlook and character we have a picture. We know next to nothing of any Chinese before him, and he lived around 500 B.C. Even of *his* life we know almost nothing, and the same is true of other great Chinese thinkers for many centuries after his time. Partly owing to his example, one respected what was old, and the individual human being was encouraged to follow tradition and not to become a unique individual whose particulars might be of interest. Again the art reflects the ethos of the culture.

Of the Greeks, however, one might expect that as from the fifth century, if not from the end of the sixth, they would have left us portraits of their poets, statesmen, and philosophers. And if the popular image of Greece were right, they surely would have done this. Of course, we do have portrait heads of Pericles, Sophocles, Socrates, Plato, and others, but they are not contemporary, and they are idealized, like the portraits of Homer. What was of interest to the Greek sculptors was the type and not the individual—Homer or Socrates as they *might* have looked.

It was only in Hellenistic times, after Alexander had conquered the Persian empire, including Egypt, Palestine, and Mesopotamia, that Greek art was transformed. Roman sculptors did many realistic portraits, again beginning in Hellenistic time and above all in the imperial period after the Roman conquest of Egypt. The most astounding portrait painting before the Renaissance are the Fayum portraits that were found in Egyptian tombs of the Roman period.

It tells us a great deal about Christianity that the European Middle Ages produced hardly any portraits. The Hegelian cliché that it was Christianity that discovered the infinite worth of every human soul is ridiculous. One only needs to recall that Christianity taught that the vast majority of humanity would be damned and tormented eternally. Much less did Christians believe in the infinite worth of each physical individual.

27

Renaissance humanism brought a new ethos. But Michelangelo, for example, was in important respects still a Platonist, concerned with ideal types rather than passing likenesses. This is a commonplace among Michelangelo scholars. But some of his predecessors and many of his contemporaries, including Raphael, Titian, Dürer, Cranach, and Holbein, painted magnificent portraits. We know how Julius II and Henry VIII, Thomas More, Erasmus, and Luther looked.

There can be no doubt of the great Renaissance painters' and sculptors' ability to capture their own image. It is doubly remarkable how rarely they did this. In the fifteenth century there is one outstanding example: Lorenzo Ghiberti included his own head on the bronze door of the Florence Baptistry, the so-called Porta del Paradiso. A quarter of a century later, Botticelli included himself in his painting "The Adoration of the Magi." Another quarter of a century later, around 1500, we find Luca Signorelli standing on the far left of his fresco "The Antichrist" in the Orvieto cathedral. Around 1510 Leonardo drew a famous self-portrait which shows him at roughly the age of sixty. Michelangelo caricatured his own features in his "Last Judgment" on a flayed skin dangling from the left hand of St. Bartholomew. He was over sixty by then, and when he looked into the mirror, he evidently did *not* feel that time was an artist.

By then a great many artists had done self-portraits,* but Dürer seems to have been the first one—and for a long time the only one—who explored, at least to some extent, what time did to his features. He did a few drawings of his own face, and three major paintings which he dated 1493 (now in the Louvre), 1498 (now in the Prado), and 1500 (in Munich). But once he was thirty he ceased painting self-portraits.

No artist before Rembrandt explored time's work on his face as he did. In an obvious way age made him less and less handsome, but he was neither vain nor a Platonist. And as we do not find Jacob or Saul at an advanced age less worthy of our interest than in their, no doubt, more handsome youth, Rembrandt found his own features endlessly interesting, and through his paintings changed our sensibility.

For him, the body was not the soul's prison—to cite the old Orphic formula that Plato quoted more than once. He did not simply express his delight in physical beauty. Nor did he merely aim to show, as others had done before him, how the face expresses the character and the soul is made flesh. What was new in his art was Rembrandt's perception of time. It is as if the people he painted did not exist in a moment, but over a lifetime, and as if Rembrandt were showing us, at least in many cases, the past of his sitters as well as their attitude toward the future.

One might wonder whether this way of putting it is not too romantic, and whether it does not read into the paintings what really is not there. One might, indeed, if it were not for the master's self-portraits. But here he clearly did something no painter had ever done before. He showed the development of a character—time's cruelty as well as her artistry.

28

Rembrandt translated the Hebrew ethos into oils. He remains a *non plus ultra*, and in a way our story comes to an end with him. But as we look back, knowing him, we find something not altogether dissimilar in the distant past.

In Egypt, during the twelfth dynasty under Sesostris III and again in the eighteenth dynasty during the so-called Amarna period, some sculptors recorded what age and suffering did to a few men and women more than three thousand years before Rembrandt's time. Looking in the opposite direction, we naturally find Rembrandt's influence in many later painters. But only one created a really comparable legacy.

Van Gogh, Dutch like Rembrandt, did all of his paintings in less than ten years, and most of them within five. One would think that a series of self-portraits done between 1886 and 1890 could not possibly show the effects of time. But this may well be the most graphic example we have of the difference between lived time and astronomical time. "A rose is a rose is a rose," said Gertrude Stein. But a rose to an aphid, a rose to a lover, and a rose to an impressionist are not the same thing. Much less is the passing of one year necessarily like the lapse of another. And few human beings have ever lived half as intensely as Vincent van Gogh did during his last five years before he shot himself, feeling unable to keep a grip on his sanity.

One only needs to see the man's paintings along with their dates to see that. The intensity is there in the pictures, and the realization that all of these hundreds of pictures were painted in such a short time adds to the overwhelming impression.

Looking at a single self-portrait of the last years, without making any comparisons, one is still addressed by a human being of such intensity that one feels and knows how he is perched in a storm between past and future. He is worlds removed from the timeless repose of Greek marble statues of the fifth century. Time is an artist but a cruel master.

---

* See, most conveniently, Ludwig Goldscheider's *Fünfhundert Selbstportraits von der Antike bis zur Gegenwart.*

II
TIME'S EFFECTS AND RESTORATION

29

Art is a response to time. It may be an escapist attempt to turn one's back to time—a triumphant act of defiance that tries to make permanent what time would destroy—or a record of time's artistry. But whatever the artist's relation to time may be, his work in turn is exposed to time.

When we focus attention on time's effects on works of art, what strikes us first is her wanton destructiveness. Even literature is not immune to that. As already noted, of Sophocles' one hundred and twenty plays only seven survive. Of the philosophy written by Greeks before Plato, only a few quotations in ancient authors have reached us. Of some of the Hebrew books mentioned in the Bible, only the names have escaped oblivion.

Architecture, sculpture, and painting have fared even worse. Nothing remains of Solomon's temple or even of the temple built after the Babylonian exile. In other parts of the world whole cultures have disappeared leaving scarcely a trace. What survives from the ages before the last two thousand years is little but a few ruins and fragments.

Of course, it is not quite fair to blame time for all that. Much of the blame falls on men who wantonly burned and broke the most glorious human creations. But time gave them their chance, and sometimes she does her ruinous work without human assistance. The claim that time is an artist is paradoxical, since what meets the eye is frequently the opposite. Yet the claim tries to call attention to what is often overlooked.

We have examples from nature in Section 1, and they could easily be multiplied. In the realm of art, ancient glass and bronze furnish uncontroversial examples. Or do they? Some people love the golden look of new bronze and prefer its luster to the subdued and subtle green patina with which time covers its sheen. The Buddhists in Bangkok like rows of shiny bronze Buddhas, while Western collectors, at least in the twentieth century, like the irregular greens and occasional blues of a good patina much better. If the point could be settled by agreeing that a work of art ought to look as much as possible as it did when it left its maker's hands, the Bangkok Buddhists would simply be right, and the lovers of patina wrong. The case for patina depends on the claim that, at least in some cases, time improves art.

This is a question of taste, and many people believe that about tastes one cannot argue: *de gustibus non disputandum*. But tastes can be acquired or outgrown. Some books that seemed good when read for the first time are found to be poor on a second reading, while others wear well, and every reading prompts new discoveries. Similarly, some works of art are flashy and look rich, though on closer inspection they turn out to be poor; others are subtle and do not yield themselves quickly, but if one lives with them one discovers how rich they are and does not tire of them.

30

The case for patina rests on its subtlety. It looks different from every angle and, for all its simplicity, offers the eye an endless variety. It is like nature. A sunset sky is not evenly golden. One looks at it and discovers how very few color words human languages have. There are no terms for most of the colors one sees in the sky, or in a patch of flowers, or foliage reflected in water.

The beauty of patina is that its subtlety is not contrived but natural. Initially, a work of art often looks artificial. It is an affront to nature, a minor outrage, and in the case of large buildings, often a monumental act of violence. Man pits his strength and his craft against nature and violates her. But time slowly restores a subtle balance. She allows nature her rights, and the beauties of art and nature blend.

An old wooden door is as good an example as glass or bronze. So is a painted wall on which the paint slowly peals. Or look at a gaudily painted boat in front of an old wall in Venice. The effect is always comparable to a reflection in water. Even a statue that invites such epithets as *kitsch*, corn, or camp, does not look so bad when it is reflected in slowly moving water.

Light often has the softening effect of time. Dimness may veil the offense and achieve enigmatic elusiveness. Or the play of light and shadow can be time's way of restoring nature's claims. The windows in the Taj furnish fine examples. Here this is clearly part of the architect's plan, and the same device is encountered in many other Muslim structures, including the Friday Mosque in Isfahan and Humayun's Tomb in Delhi. San Pietro in Assisi offers a more unusual variation.

31

Fragments have much in common with patina.

> Looking at a statue that is broken, we feel the texture of the stone, vibrant with life and with uncertainty. It makes us muse, whether we think of it that way or not, over human possibilities. . . . what remains calls into question our previous notions, and the artist's, too; and something new is seen. . . . There is something offensive in a sculpture: it is an attack on the human body and face which after all are not simply stone—hard, inflexible, smooth.
>
> A face and body are never finished while they are alive. That is why Michelangelo's unfinished sculptures are more alive than those he finished.
>
> Look at two scluptures . . . In both the tip of the nose is broken. Nothing worthwhile seems lost. Rather the breach reminds us unconsciously that these are not passing likenesses which some such flaw might well reduce to worthlessness. These faces are not functional. That little loss is liberating. Unobtrusively it does what modern artists often go to such great lengths to do. The break frees us from bondage to the facts and is a triumph of the spirit.*

Time takes her revenge on the sculptor's impiety, of which Moses and Muhammad were so keenly aware, and destroys the effigy without impairing the art. It sets free the spirit imprisoned in stone and restores possibility, freedom, and imagination.

Can one convince a person who does not see it that way that a man or woman in marble is an offense? Hardly by abstract argument, if at all. In the Accademia in Florence, many people walk past the unfinished sculptures of Michelangelo on their way to his huge, unimpaired "David," which is the featured piece at the end of the hall. They are told that he is the marvel of this collection, and they believe it. Is it simply claim against claim, taste against taste?

Perhaps not. One might begin by looking at all the Michelangelos in that room, including the "David" as well as the unfinished "Captives." One might contemplate them with the respect due to a supreme artist and an enormously interesting human being. For some people nothing more is needed than simply to look for some time, to stay in that room for an hour, walking from this piece to that and looking again and again from different angles. Others may want to be told what the project was of which some of these sculptures were meant at one time to be a part; still others, how these unfinished pieces have changed the course of art and man's self-understanding.

What is wanted is knowledge, not of the sculptor's life but of the reasons for which many sensitive people have admired these sculptures, and the reasons for which the David has been admired. That the David is a fine sculpture is undeniable, but its fame rests rather unsubtly on its enormous size. To that one can add the story that this block of marble had seemed unusable, and that Michelangelo surprised everyone by making this David out of it. If it were half its size, it would not attract half the crowds that now come to see it.

The incongruity between the subject and the execution seems to escape notice. This is supposed to be David who with his slingshot brought low the giant Goliath. But what Michelangelo gave us was a David so huge that one must forget about Goliath. There is the slingshot, but why should this David need it? One blow with his enormous right hand would topple anyone. Surely, this is not David; these are not the hands that played the harp before Saul; this is not the face of the poet; there is nothing here of the Biblical David, least of all of the old king who cried out: "My son Absalom, my son, my son Absalom! Would I had died for you, Absalom, my son, my son!"

* Kaufmann, *Critique of Religion and Philosophy* (1958), section 87.

On the ceiling of the Sistine chapel Michelangelo showed that he could paint not only the Biblical Jeremiah and Jonah but even—impossible as it seems—God creating the sun, and the creation of man. But surely "David" is the one work by Michelangelo that has been overrated.

In its defense one might say that it represents the triumph of Renaissance humanism. In some ways the bronze Colleoni on horseback, done by Verocchio in the 1480s, is a more striking symbol of defiant individualism. Yet David has no horse, no clothes even. Here is naked man confronting the world with youthful strength. Still, when compared with the captives he seems to lack depth; the spirit that grows through suffering; humanity.

In the unfinished sculptures at the Accademia we find these qualities in the highest degree. Here are men struggling with torment, feeling the hopeless agony of being human, and striving against impossible odds. Here the artist does time's work, without waiting for her, and creates something akin to fragments.

32

Michelangelo did not invent the *non finito*, the unfinished work. The first great example was the tower of Babel. When *Du*, a beautiful illustrated monthly published in Switzerland, devoted an issue to the *non finito* (in April 1959), it opened with a color reproduction of Bruegel's "Tower of Babel" in Vienna. The articles that followed included "Incomplete Representations on the Walls of Egyptian Temples and Tombs," "Uncompleted Greek Sculptures," and "Beauvais: The Uncompletable Cathedral."

Michelangelo's unfinished sculptures bear witness of his genius. *He knew when to stop.* This was an extraordinary discovery. We need not suppose that it was all planned that way. It is sufficient that the old sculptor stopped, in work after work, when he had achieved an expressiveness that was *non plus ultra*, when nothing was to be gained and a great deal to be lost by further work. We need not imagine that he felt exultant as he saw that his sculptures could not be improved any more. On the contrary. What these marbles, including his late Pietàs, express is precisely the limits of human power, the finality of suffering, frustration, and disappointment—the burden of man's lot.

None of the sculptors who sought to imitate Michelangelo's *non finiti* achieved anything even remotely comparable, with the possible exception of Auguste Rodin, well over three hundred years later. But even in the work of Rodin, Michelangelo's torment became a polished manner.

In literature the fragment became a genre during the age of romanticism. What set the stage for that was *Faust: A Fragment*, published in 1790 by Goethe, who had as much trouble with *Faust* as Michelangelo had with the sepulcher of Pope Julius II. Goethe actually had a manuscript in which the first part of the tragedy was completed, but he was unhappy with the style of its stark conclusion; and having been unable for about fifteen years to satisfy his own standards, and having no more than Part One in any case, he published the fragment.

As in Michelangelo's case, this form was profoundly appropriate. For Faust is the embodiment of ceaseless and insatiable striving.

Quickly, the German romantics developed a predilection for fragments. And after Goethe had set this example, German philosophers from Hegel down to Heidegger did not consider it in the least embarrassing to publish Part One of a major work and then let it go at that.

What, it may be objected, does all this have to do with time? What is finished is apt to suggest that time stands still, that change has been arrested or is in some sense unreal, inferior, and can be transcended. Michelangelo's unfinished works suggest, on the contrary, that we cannot transcend time, change, and suffering, and that man's perfection, such as it is, consists in striving. In Goethe's *Faust* this ethos is spelled out expressly.

33

Ex ungue leonem, "From the claw you can infer the lion," is a Latin adage that can be traced back all the way to a fifth-century Greek poet. To this one might add, "and from the hoof the ass."

Not every fragment suggests a master. The torso of the Venus of Cyrene in the Terme Museum in Rome (the first picture in the final color section below) is a *non plus ultra*. One feels grateful that she has no head or arms. This feeling is strengthened by knowledge of more complete statues of this type (see the next picture). To be sure, even time could not have turned the second Venus into a great work of art. But torsos like that of the Venus of Cyrene change our perception of finished sculptures. We ask ourselves what amputations might improve them. The third picture in this sequence is interesting to consider in this perspective. Here, for once, time has not gone far enough.

Michelangelo said that a good statue can be rolled down a hillside without breaking. It was a prophetic indictment of the baroque extravagance which was, against his will, inspired by him: florid and pompous works that abound in theatrical gestures. His saying—like the Venus of Cyrene —brings to mind a couplet by a German poet who lived during the high tide of the Baroque, Johannes Scheffler (1624–1677), better known as Angelus Silesius.

> Become essential, man! For when the world is done,
> All accidents will drop, the essence will remain.
>
> *Mensch, werde wesentlich; denn wenn die Welt vergeht,*
> *So fällt der Zufall weg, das Wesen, das besteht.*

It is a crucial aspect of time's artistry that it often removes what is not essential. For most artists do not know when to stop. Less can be more.

34

Ruins are architectural torsos and often have a rich patina. Even an ugly building can become an acceptable ruin. The Kaiser Wilhelm Memorial Church at the end of the Kurfürstendamm in Berlin was an eyesore. What was left of it after the bombings of World War II was given a very appropriately shortened name and serves well as the Memorial Church, bringing to mind the war rather than the Kaiser.

Ruins in busy cities always serve a memorial function. They remind us of time, change, and death. Today, Lopburi is a provincial town in Thailand, but it was once a center of the Khmer civilization and gave its name to countless Buddhas in stone and bronze. The ruins of its medieval temples that still stand there in green surroundings, only a few yards away from the bustling streets, evoke some sense of the passage of time. They give modern man some perspective on his own culture, and invite some reflection on how we are spending our lives. To some people rushing past or sitting in their shadow they may well say in one language or another:

Mensch, werde wesentlich!

Most of the best-known and best-loved ruins are to be found some distance away from modern cities, in the desert or among trees and wild flowers. In such settings it is even more obvious that time has restored a balance, that nature is reclaiming her rights, and that the beauties of art and nature are blending. Even those who do not care to go to church are apt to be moved by Gothic ruins without a roof, with the sun shining on red poppies where the altar once stood. Let those who do like to go to church build theirs elsewhere and not destroy time's work!

35

The problem may seem to be very different when the building in ruins is believed to have been a masterpiece. Then there is apt to be a great hue and cry that it should be rebuilt in its ancient or medieval glory, just as it was when it was it was new. And this strange suggestion is passed off as a plea for "restoration."

If the building was once a church or a temple, the point is not that some people now want to worship there. The plea is that such work should be done for the love of art, and vast sums are raised for the so-called restoration of major artistic sites.

What originates as an interest in art and in the historical past, and often gains the support of governments, or foundations and corporations who are persuaded that they should give money for culture, has often turned out to be devoid of historical and aesthetic sense and destructive of culture.

Consider some easy cases first. Sir Arthur Evans dug up and "restored" Knossos in Crete. On one wall of the palace he found fragments of a second millennium fresco showing parts of a blue figure. He had a painter complete the picture. In the second edition (1927) of *Die Kunst der Antike* (*Hellas und Rom*) in the prestigious and beautiful multi-volume *Propyläen-Kunstgeschichte*, the first color plate still shows us the amazing result: a blue boy picking flowers. A little later it was discovered that Sir Arthur had made a mistake. The old fresco had portrayed a monkey. But it was too late.

Anyone who admires Franz Marc's expressionist painting of blue horses will not take offense at the idea of a blue boy, though it is amusing to see the latest styles projected so quickly into the second millennium B.C. Anyone who does not happen to like the Swiss painter Ferdinand Hodler will be distressed to find that so many old Cretan paintings bring him to mind. And this distress is not likely to be lessened by the discovery that the painter who worked for Sir Arthur was Swiss.

It seems elementary that no painting should have been completed, "restored to its original state," or tampered with *in situ*. While it is legitimate and extremely interesting to try to guess how the paintings may have looked more than three thousand years ago, the place for such efforts is clearly a nearby museum, not the ancient wall, and least of all the area in which patches of the old painting had still survived.

Even black and white photographs show at a glance how gaudy Sir Arthur's restorations are and how little feeling they show for the ancient site. The lack of reverence, of historical sense, and of aesthetic sensitivity is extreme in this case. Why did all this have to be done in Knossos, why not in some archaeological Disneyland a few miles away?

36

The Egyptian government has given the United States a complete ancient temple that is to be assembled in the Metropolitan Museum of Art in New York City. This is to show its gratitude for American generosity in helping to move the temple of Abu Simble, with its gigantic statues of Ramses II overlooking the Nile, from its ancient site, which was flooded when the Aswan Dam was built.

In an age in which such things can be done no excuse remains for the barbarous "restorations" that are still carried out in ever so many places. Those who want to see everything as it was when new should leave the ruins for those who appreciate them and do their reconstructions somewhere else. There is almost always plenty of space nearby, not to speak of the parks of London and Paris or the wastelands of Florida.

Illogically, it is argued that people who travel to faraway places want to see what can be seen only there. But what is truly unique in those places is time's work on remnants of the distant past.

Restorers are undertakers who refuse to countenance time, change, and death. They have no respect for age, for texture, for patina. They are unwitting enemies of history and of culture.

37

Paintings pose special problems that should be considered before we turn to the crucial distinction between restoration and conservation of buildings. Actually, the restoration of paintings has been discussed far more in the past than has that of buildings. Goya already was asked for his verdict on a restorer of paintings, and on January 2, 1801, sent a report to Don Pedro Cevallos. After examining the pictures cleaned and retouched, with an eye to "the improvement or damage the paintings may have undergone by such treatment," and after also examining "the method and the quality of the ingredients that he uses to brighten them," Goya passed the following verdict:

> It would be hard for me to exaggerate to your Excellency the discordant effect that my comparison of the retouched parts with those unretouched caused in me, because in those retouched parts the verve and vigor of the paintings and the mastery of the original delicate and knowing touches still preserved in them had disappeared and been destroyed completely. With my innate frankness, which is moved by my feeling, I did not conceal from him how bad all this seemed to me. Immediately thereafter others were shown to me, and all of them equally deteriorated and corrupted in the eyes of professors and true intellects, because in addition to the invariable fact that the more one retouches paintings on the pretext of preserving them the more they are destroyed, and that even the original artists, if they were alive

now, could not retouch them perfectly because of the aged tone given the colors by time, who is also a painter, according to the maxim and observation of the learned, it is not easy to retain the instantaneous and fleeting intent of the imagination and harmony of the whole that was attempted in the first painting, so that the retouches of any subsequent variation might not have adverse effect.

38

The notion that time is a painter was developed earlier in a little sketch by Joseph Addison in *The Spectator* of June 5, 1711:

> . . . I dreamt that I was admitted into a long spacious gallery which had one side covered with pieces of all the famous painters who are now living, and the other with the works of the greatest masters that are dead.
>
> On the side of the living, I saw several persons busy in drawing, colouring, and designing; on the side of the dead painters, I could not discover more than one person at work, who was exceeding slow in his motions, and wonderfully nice in his touches. . . .
>
> Observing an old man (who was the same person I before mentioned, as the only artist that was at work on this side of the gallery) creeping up and down from one picture to another, and retouching all the fine pieces that stood before me, I could not but be very attentive to all his motions. I found his pencil was so very light, that it worked imperceptibly, and, after a thousand touches, scarce produced any visible effect in the picture on which he was employed. However, as he busied himself incessantly, and repeated touch after touch, without rest or intermission he wore off insensibly every little disagreeable gloss that hung upon a figure. He also added such a beautiful brown to the shades, and mellowness to the colours, that he made every picture appear more perfect than when it came fresh from the master's pencil. I could not forbear looking upon the face of this ancient workman, and immediately, by the long lock of hair upon his forehead, discovered him to be *Time*.

39

In 1816 Goethe was consulted about the restoration of paintings in Dresden, and on April 9 submitted a report which he signed together with his friend Heinrich Meyer, who was a painter and art historian. They condemned all attempts "to make good old paintings appear as if they were new, for intensive washing and alleged cleaning of the light parts removes the so-called patina," and along with that the color balance and harmony of the whole. Only "when a painting had become totally unenjoyable and entirely black in the shadows" did they consider it right to lighten the shadows because "only the areas that have become entirely dark and unclear are disagreeable and disturbing for the informed viewer." They agreed that "cleaning and restoring should be considered only as a last resort and be risked only when paintings have become totally unenjoyable."[*]

It might be supposed that since the days of Goya and Goethe restoration techniques have improved so much that their views are dated. Nothing could be further from the truth. In 1956 Jakob Rosenberg said in *The Art Quarterly*:

> In the national Gallery in London it was possible this year to see side by side two Rembrandt paintings representing the same old lady (Margaretha Trip) at about the same time, one of them thoroughly cleaned and showing the kind of harsh effect just mentioned, with an exaggerated contrast of light and dark and a loss of transitions and overtones; the other one uncleaned, *i.e.*, with its old varnish which, however, is still sufficiently transparent and shows just a slight yellowing. I have not the slightest doubt that Rembrandt himself would have preferred the effect of the uncleaned picture. One has only to look at the master's etchings and drawings to be convinced how vital to him was the total tonal harmony, and consequently the transitional tones as well as the overtones. And I doubt whether it was any different with Titian, with Ruben, with Velasquez—in short, with all the great painters. (p. 389)

In four immensely erudite articles, two by each, E. H. Gombrich and Otto Kurz quoted and cited dozens of authors from Pliny to Rosenberg who had expressed themselves along the same lines.[**] "Guido Reni said that while the works of other painters lose something in the course of time, his own will, by yellowing, acquire a new lustre. The patina will bring out the mutual harmony of colours."[***] He reckoned in advance with time's effects, and many other artists made

[*] Printed in *Jahrbücher für Kunstwissenschaft*, IV (1871), p. 261.
[**] *The Burlington Magazine*, 1962 and 1963.
[***] Quoted by Gombrich, 1962, p. 58.

the same point, down to Renoir who said that a painting depends "not only on the elements of color and design but also on the element of time. For the patina of time is no vain word; but the main thing is that a work supports this patina. And only remarkable works can support it."*

40

Even as the restorations at Knossos reflected the taste—or the lack of tasfe—of the restorer and his generation, the current insistence on ever more restorations reflects the taste—or the lack of taste—of our own age. To get people into museums, one has to show them something new every time. The same old pictures will no longer do. Some people, of course, go to a museum precisely in order to see some paintings that are like old friends for them. But some museum directors are not concerned about people like that. They come anyway. But to bring in masses of people and get media attention, one has to show new things. The old masters get cleaned, the media report it as news, and masses of people come and admire the gaudy colors.

The craving for what is new is satisfied best by what also *looks* new. And even some of the finest museums cater to this vulgarity. The approach is essentially that of department stores advertising a sale and hoping that on their way to the sale area masses of people will buy things that are not on sale. One advertises a show of Scythian gold, knowing very well that few people care about the Scythians, but that gold will prove an enormous attraction; and gold always looks like new. And one hopes very reasonably that of the masses who come to see all that gold, some will also look at a few of the other new exhibits. Thus the museum comes to see itself as show biz.

Still, the museums are far better than the theaters. Theaters depend for their survival on drawing crowds and devote themselves almost exclusively to what is new. On the rare occasions when they stage an old play, it is almost always refurbished. Plays that involve translations from other languages are often changed beyond recognition to make them contemporary. And Shakespeare in New York requires some gimmick, some bold new angle. See Schofield play Lear without passion! Or Burton as Hamlet without any touch of melancholy!

The theater, too, is not for those who want to see some very old friends. The purveyors of culture who get government grants do not aim to provide us with some perspective on our time, our culture or lack of culture, our beliefs and ways of life. Even university theaters no longer aim to do that, but prefer to show what is new, very new. Everything has to be contemporary.

This is not the way restorations are usually discussed. Naturally, that debate, too, has been blind to the historical setting of the whole problem. But one needs some perspective, some sense of the nature of our time and its place in long-term developments, to understand what is at stake.

Our culture no longer respects age. In the cities, old people are mugged for sport. On buses, and even in college lecture halls when a guest speaker draws a large crowd, young men sit while old men and women stand. Even students at the best colleges have not been told that this is uncouth, and if you quoted them the Bible's "Before a grey head rise," they would not know what you were talking about. They have never read Leviticus 19, which also introduced such ideas as "You shall love your neighbor as yourself" and "The stranger who sojourns with you shall be to you as the native among you, and you shall love him as yourself." Even professors no longer have time to read things as old as that. They are under far too much pressure to read the most recent twaddle.

How, then, can one expect people to love and admire old paintings unless they are cleaned first and made to look new? That is the problem, and all the talk about the masters' intentions is merely a smokescreen. Goyas are cleaned along with all the rest.

41

Apologists for the restoration of *buildings* often employ another argument. We need to do something, or the building will collapse altogether. Rarely is that true.

Time is an artist—who does not know when to stop. The most beautiful torso does not suffice her. She does not put down her chisel leaving well enough alone. She keeps at her work relentlessly until nothing is left. She is the great Creator and Destroyer who destroys all she creates, slowly but ceaselessly.

From a purely aesthetic point of view it makes little sense to stay her hand, or even to try to slow her down. When the Houses of Parliament burned, the greatest painter in England did not try to put out the fire, but painted the glorious spectacle.

---

* Kurz, 1963, p. 96, quotes the original French.

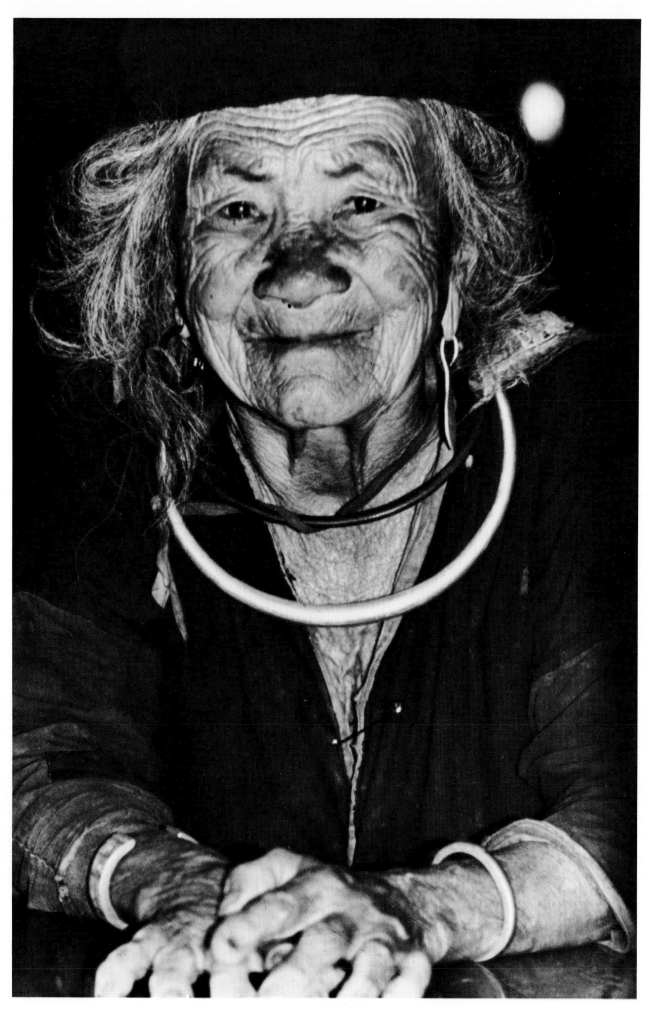

Meow woman, said to be 100 years old. Thailand, 1971

Stonehenge, England. 1952

Varenna,
Italy. 1975

Acropolis, Athens. 1956

Knossos, Crete. 1956

Conference on Aging, Cleveland. 1975

The only good reason for conservation is not aesthetic, but a concern for posterity. This involves seeing oneself as a mediator between past and present and a link in a momentous tradition. In that case one feels a responsibility both to one's ancestors and one's descendants, and wants to make sure that the works which shaped us and our parents and parents' parents will still be there to shape our children's children. That would make sense if it were not rendered absurd by our whole educational system and the facts mentioned in the last section. I am not against conservation that is done in this spirit, but to do that honestly would require far-reaching changes throughout our culture, and not by any means only Western culture.

Actually, multi-million dollar projects are rarely content to do what is needed to prevent total destruction. Talk of that is for the most part mere propaganda, and what is attempted in fact is restoration, making things look like new, or rather a half-hearted, inconsistent, indefensible program of restoration.

The technical aspects of such projects—what precisely is needed or feasible in a given case —must be debated on the basis of research into local conditions and will involve engineering questions. But photographs can show what is being done and should not be done.

42

Borobodur in Central Java is one of the grandest religious sites in all the world. It is the largest Buddhist structure anywhere, exceeding in sheer size the Bayon at Angkor. Indeed, the most nearly comparable sites are those at Angkor and the Hindu temples at Prambanan, also in central Java.

The best individual sculptures from Angkor, including many from the Bayon, have an expressiveness and an enigmatic spirituality not matched by the countless Buddhas of Borobodur. The difference is not merely one of degree, it is overwhelming. The statues at Borobodur are fine, but the carving does not approximate the haunting quality of the Bayon Buddhas and bodhisattvas.

At Angkor, this haunting and enigmatic quality is reinforced by the jungle setting and the effects of time. (See picture 44.) The French did a wonderfully restrained job of conservation. Some places they left as they found them in the jungle, doing only the minimum work required to prevent collapse; and elsewhere, too, they did not feel called upon to clean everything and remove the patina of the centuries. The balance of nature and art here is unforgettable.

At Prambanan, which anyone in a hurry can see on the same day as Borobodur, the tower of the main Shiva temple had toppled,* and the Dutch restored it. (See pictures 41 and 42.) It is a measure of their success that anyone who has not looked at old books and found pictures of Prambanan before restoration would never suspect that the tower was rebuilt in the twentieth century. It does not look new. It blends with the landscape, and enough ruins are left all around it to give one a sense of the place, and of time.

Something like this has been done at Selinunte in Sicily. A few ancient Greek columns reassembled and even one temple partially rebuilt with the old stones can make a marvelous contrast with a vast field of ruins. What we see in such cases is not an absurd attempt to turn back the clock and show us everything as it once was, on the offensive assumption that we can appreciate only what has a new look, and that what looks old cannot be beautiful. On the contrary, if one has not been there before the work was done, or seen pictures take before (which, given the vastness of the site, might not even appear to be conclusive), one might never think that anything had been restored. The poignancy of the destruction of Selinunte has only been heightened and time's work underscored, not undone.

Before embarking on any project of conservation or restoration, one must discover the genius of the place and then lend every effort to letting it speak in its own language and tone of voice. To silence it and replace it with a modern idiom is cultural genocide.

43

Borobodur is a holy mountain, but as unlike Mount Sinai as can be. Java has huge volcanoes that rise from the plains to impressive heights, and one sees some of them on the way to Borobodur, and again from the top. But Borobodur does not look in the least like them. In fact, one does not really see the holy mountain as a mountain, a part of nature, because the whole of it has been transformed into a stupa. One ascends on steep, regular steps, but not straight to the top. The

* Karl With, *Java* (1920), photographs 62 and 63.

ascent is broken by terraces that gird the whole mountain and are adorned with continuous friezes that show us scenes from the previous lives of the Buddha.

The quality of the friezes is fine but not startling, and the effect of so many reliefs is numbing. One feels, and is clearly meant to feel, caught in the seemingly endless repetition of so many cycles. What makes it bearable is time's work on the sculptures and on the occasional statues of Buddha. The color of the stone is not the same everywhere; there are breaks; and lichen and moss reassert the rights of nature.

The statues of the Buddha, too, have been freed by time from their monotony. If it were not for that, one would soon come to feel that if you have seen a few, you have seen them all. For terrace upon terrace offers essentially the same sights, except for the landscape in the background, which changes as one climbs higher.

Emerging from these narrow paths and steps to the highest terrace is exhilarating. The experience suggests the Buddha's triumph over seemingly endless rebirths and the liberation of final enlightenment.

Nothing quite like this had ever been achieved by art. The pyramids were not meant to be climbed; and if you nevertheless scale the tallest one, you are not distracted by any sculptures, friezes, or horizontal paths. Nor was anything like this attempted at Angkor. Borobodur, crowned by a large central stupa and by many smaller ones that contain statues of the Buddha, is unique.

Now it could be argued that on my own account time's work obstructs the message and that everything should be cleaned for the sake, not of lovers of art, but of Buddhist pilgrims. One would then say that beauty has to be sacrificed to salvation; culture and art, to Nirvana. In fact, this has not been the reasoning behind the vast restoration of the 1970s. The government of Indonesia and the people of Java are Muslims, and of the few Buddhists who come there most are Japanese taking endless snapshots of each other. Even apart from that, this argument is, of course, ridiculous. The very idea of collecting millions upon millions of dollars from governments and corporations to clean these friezes and statues for the sake of Nirvana is grotesque.

If so, it may be countered, the very idea of building Borobodur was grotesque in the first place. To be sure, the enterprise had its ironical side even then. The Buddha taught men and women not to rely on means like these, and in early Buddhism sculpture and painting were strictly forbidden. After that ban was broken, the prohibition of images of the Buddha persisted for a long time. The countless Buddhas of Borobodur are fraught with world-historical irony. They represent a triumph of Mahayana Buddhism, which developed centuries after the Buddha's time, and the holy mountain reveals the spirit of the age in which it was built, one thousand two hundred years ago, around 800.

It belongs to that time, that moment in history, when Mahayana Buddhism flourished in central Java. Cleaning the statues and friezes cannot restore that moment or bring back the people who climbed the mountain in those days. Today's Buddhists are different, very different from the Javanese Buddhists of that time. To obliterate as best we can the sense of distance created by patina, lichen, and cracks does not help anyone's spiritual quest. We need the contrast between this world of old stones and the world we normally live in. We need all the help that time's work provides.

44

The restoration does not enhance the sense of liberation as one reaches the final terrace. On the contrary, it has all but destroyed that experience. The last three pages of the second color section above show the transformation of the highest level. First you see, on facing pages, the cleaning process and the appearance of the terrace before the cleaning. In the left picture one sees the contrast between the parts that have been scrubbed and those still bearing witness to time's artistry. The final page shows the end result. The supreme terrace has been turned into a kind of Disneyland where tourists climb over the statues and have their pictures taken with one arm around the Buddha. The sense of distance has been erased, the aura is gone.

If one had the good fortune of making the ascent before the restoration, even as recently as 1971, one can recall the sublimity of the place, the numinous atmosphere as one seemed to be above the world and beheld a great volcano beyond the stupas and an enigmatically discolored Buddha image. Now the symbiosis of art and nature is broken, the ruins that are more than a thousand years old have a new, artificial, lifeless look that brings to mind plastic. The spirit has been driven out of the stone, and what is left is dead.

The place is full of life, you may object, full of tourists enjoying their outing. Are people not coming from far away to see the new Borobodur? No longer is it a lonely place, known to a few with an interest in art and religion. There probably has not been so much life here since the ninth century. Look at the thriving stalls at the bottom where dozens of hawkers make a good living. Isn't it worth the vast expense? Tourism may yet pay for it.

How long will it take before one discovers that the masses of utterly uncomprehending tourists are ruining what restoration has not destroyed? How long before the old Buddhas are taken away to the relative safety of a museum and reproductions installed in their places?

Then it will be too late to suggest that the money misspent here might have been used instead to create to copy of Borobodur somewhere else. By charging admission one might have recouped the vast expense, and the original site might then have been saved.

Whatever work was really needed to conserve it should have been kept to the minimum, as in some parts of Angkor. But the people in a position to spend vast sums of money on culture do not know for the most part what they are doing.

45

Near the end of this book there is a color photograph of a clay pot made in Abraham's time and found somewhere near Jerusalem. Its patina speaks for itself, and so do the five pages of photographs of Jerusalem. Actually, all of the pictures tell their own story, and after simply turning the pages slowly and looking at them repeatedly some readers will understand not only how time is an artist but also what happens when people forget it.

Still, words can add another dimension. The pictures before the Middle Bronze Age pot show some unrestored details of the Haram al-Sharif, the ancient Jewish temple square. The monuments are all Muslim, and the Haram is one of the glories of Islamic art and as beautiful as any man-made place in the world. It is quite unlike the Acropolis, Angkor, Borobodur, or the pyramids; it is unique.

That time has much to do with its charisma is obvious. David, the Biblical David, stood here, and Solomon built his temple here. During the exile, Ezekiel kept alive his people's longing for a return to this spot, and when they returned they rebuilt the temple. Herod beautified the whole site, and the lower part of the Western Wall dates from his time. So does the design of the platform on which the Dome of the Rock now stands along with the smaller Dome of the Chain and a few much smaller dependent structures.

The Dome of the Rock was completed in 691 A.D., sixty years after the death of Muhammad, but is much closer in time to us than it is to David and Solomon. It took courage to build in that setting, which the Christians had spurned. The Romans had not left one stone on another. There were plenty of stones then, but the Dome of the Rock was not simply the best that could be built with the means at hand; it is the oldest mosque in the world and one of the finest buildings anywhere. The design, the proportions, and its relation to the surrounding square and other structures are all perfect. And restoration has impaired none of this.

A comparison of the Dome of the Chain, still unrestored in 1975, with the Dome of the Rock after restoration is nevertheless striking. It shows the difference between the old tiles and the new ones. Of course, the old tiles did not go back to the seventh century, only to the sixteenth. Still, four hundred years sufficed to give them a patina, and there was nothing gaudy about the Dome of the Chain.

By 1977, the Dome of the Chain had been stripped of its tiles, and this "restoration" project was supervised by a Palestinian Arab who, unlike all of his Muslim predecessors in this area, kept careful track of every tile. Thus we know how a great many tiles had been damaged before the job was begun, and that as many of the undamaged ones were broken during the process of removal. That sort of information is rarely mentioned in discussions of restorations, but it would not greatly influence those who favor the restorations on the Haram.

In the nineteen twenties, a Turk was asked to restore the Al Aqsa mosque on the Haram, and a large sketch of what he proposed to do hangs in the office of his successors in the Al Aqsa. Instead of spending a great deal of time and money on the old mosque, he wished to erect alongside it, in the east, a far more splendid mosque that would exceed in height not only the Al Aqsa but even the nearby Dome of the Rock which stands on a much higher platform, reached by alternate imposing stairways of more than twenty steps. The framed sketch in which the main dome rises above lesser supporting domes immediately brings to mind St. Sophia in Istanbul, which evidently was for the Turk the paradigm of a great mosque. The sketch also includes a minaret immediately to the east of the new mosque, as slim as the minarets of St. Sophia and quite unlike the minarets on the fringes of the Haram. There is no minaret in the square itself, and the extraordinary beauty of the area and the sense of peace it conveys are due in some measure to the absence of crowding and of tall structures.

Kamal Eddin Bey's proposal was rejected, but the projects of his successors have to be understood against this background, and not merely as restorations. In the spring of 1977, the chief engineer still faced that picture when he sat at his desk, and he punctuated his discussion of the restoration done under his own supervision with the refrain, "Of course, this is not the old Aqsa mosque; this is a new Aqsa mosque." He had fought hard to keep the Crusader chapel on the east side of the mosque, against objections that it looked old and that it was of Christian origin. He felt that it was part of the history of the mosque and worth keeping. As the pictures show, however, it looks like a quotation from Shakespeare in the midst of poor prose. Its artistic merit is not so singular, but compared to its texture and patina the new wall to the right and left of it looks like plastic.

Two pictures show a Byzantine capital with a rich patina on the ground outside, discarded, and the new columns being carved inside the Al Aqsa. The Italian marble columns donated by Mussolini have been retained.

The ceiling in the Dome of the Rock had been redone by the Mamluks after the Crusades, but the Crusaders' iron screen around the large irregular rock in the center, which antedates David and Solomon, was largely destroyed in the nineteen fifties when the whole building was taken apart and rebuilt by Egyptian engineers and architects. In the Dome of the Rock almost everything from the tiles outside to the rugs inside is new now and looks it.

The notion that brand-new rugs are more beautiful than old oriental carpets is worthy of a character in Molière. It meets the eye that the new rugs and tiles are shinier than the old ones they replaced, and yet they seem much less alive. The designs formed by the old tiles were often asymmetrical, like the designs on old oriental rugs. The restorers made the designs perfectly symmetrical —like machine-made wallpaper. And to top it all off in style, the modest black dome was replaced with a golden covering made of an aluminum alloy that looks *nouveau riche*.

Along with the silver dome of the new Al Aqsa mosque, this golden dome with its inorganic and lifeless look is visible everywhere, as blinding in the sun as the chrome on a car. Neither of them will ever acquire any patina. But the Egyptian engineers, coming from a country where it hardly ever rains, had no adequate conception of how one must build in Jerusalem, which has a rainy season, and in a few years something will have to be done about the two domes. It is not impossible that at that time they will be improved aesthetically, too.

When the restorers come from another country and climate, and are rooted in different traditions, it stands to reason that their work will have an inorganic quality and will be alien to what they profess to restore. Moreover, many restorers really see themselves as improvers. The Egyptians, for example, meant to move the Dome of the Chain to another location, where they thought it would look better than it did so close to the Dome of the Rock, and they actually completed the work on the new base for it in the northeast corner of the platform.

As Goya said, "even the original artists, if they were alive now, could not retouch [their own works] perfectly because of the aged tone given the colors by time, who is also a painter," and "it is not easy to retain the instantaneous and fleeting intent of the imagination and harmony of the whole that was attempted in the first painting." The restorer is typically an alien, even when he does not come from another country and climate. He is formed by another time and different traditions, and he has not lived in the soul of the original artist. As a rule, it has never even occurred to him and to those who admire his work that the artist himself could not have recaptured his original inspiration a mere ten years later. Poets and painters, sculptors and composers know how dangerous it is to touch a work done at an earlier stage of one's own life, and how it is almost always preferable to let it stand with its imperfections instead of trying to improve it from the vantage point of a necessarily different taste. But artists and restorers have different perceptions of time. And when that is ignored, irrevocable harm is done.

EPILOGUE: OLD IS BEAUTIFUL

46

Perhaps the attack on time and its work has passed its peak, and we are on the threshold of a new sensibility. There is little to indicate that people are becoming more historically minded. On the contrary, the concern with the future far exceeds interest in the past, and the cult of youth and the lust for novelty have not abated. Yet there are three small clouds on the horizon, each of them no bigger than a human hand—and one cloud like that sufficed Elijah to predict the end of a long drought.

The first hopeful sign is that voices of protest are being heard. This in itself is not new, and Goya and Goethe did not stem the tide of barbarous restorations, nor did Gombrich and Kurz. But recently the restorers have gone so far that it seems possible that a reaction will set in.

On New Year's Day 1977, *The New York Times* ran a story under the headline, "Optics at Chartres Reported Ruined." The point was that three of the most celebrated and venerated stained-glass windows in the world have been irretrievably "altered by cleaning and conservation." French artists led the protest, but were at first pooh-poohed by the officials in charge: "Who would dream of taking the word of an artist" seriously? But some scientists corroborated what the artists had seen with their eyes, and thereupon it was reported, "the Ministry of Culture has ordered all stained-glass restoration to be suspended, at Chartres and elsewhere."

At the end of January, *Time* magazine ran a long story on "Chartres: Through a Glass Darkly," and joined to it another, "Acropolis: Threat of Destruction." The Greeks, it reported, are now planning to remove the sculptures that remain on the Acropolis in Athens to an as yet unbuilt museum, replacing them "with fiber-glass replicas." All the ancient columns will require extensive "restoration." The story ended: "The idea of a Parthenon 'restored' with fiber-glass replicas, girdled by lines of tourists trudging along a viewing ramp, may be depressing, but it also may be better than no Parthenon at all."

Time's concern may have been prompted in part by Ada Louise Huxtable's eloquent editorial in *The New York Times* of January 17, 1977. Under the heading, "Alms for the Acropolis," she said, "Pilgrimages have not been made for 25 centuries to see the marbles of molded fiberglass. The sculpture will never be experienced properly, as it was conceived, again. Next—a plastic Parthenon?" Even Huxtable ignored the artistry of time when she suggested that in the 1960s the sculpture could be experienced "as it was conceived," and that this alone is the proper way of seeing it. Originally, the sculptures were painted, and the Acropolis may well have looked *nouveau riche* to Greek visitors on the eve of the Peloponnesian War. It was time that gave the Acropolis the aspect we loved. But I applaud Huxtable's conclusion: "It is the peculiar arrogance of money and technology to believe that a civilization can be put back together again."

How can one find hope in words like these? The problems have clearly become worse than ever. So have the extravagances of the believers in restoration. It is not altogether impossible that people with some understanding of culture may rally in this hour of unprecedented need. I doubt that they would be heard, if it were not for the other two clouds on the horizon.

47

The second hopeful sign is that in many of the wealthiest countries the birth rate has declined sharply. It is alarming, of course, that the birth rate has not declined similarly in most poor countries, and that in the well-to-do countries, too, the poor and uneducated keep multiplying at a faster rate than others. What, then, inspires hope?

For a long time, the people who were economically better off and more educated got younger and younger, and respect for the old and for what is old declined. Now this trend has been broken. The number as well as the percentage of the old will rise steeply, and it is to be hoped that in democratic countries they will not only demand respect but get it. If so, attitudes toward age and time may change. The era of contempt and lack of interest may draw to a rapid close.

Finally, women are at long last insisting on the respect that is due them. On the face of it, this may have nothing to do with time. But in fact, this is the most important of the three small clouds.

It is a puzzle that old women have been treated so wretchedly in so many cultures. I am not sure that this puzzle has ever been solved. The key to it is probably man's horror of time, and his reluctance to admit that the passage of time is irrevocable. One refused to admit that death was irrevocable and fantasized about life after death. That during much of their lives women tell time by bleeding periodically was mysterious and inconvenient, but bearable. From a man's point of view, it was not that different from other rhythms in nature. What was felt to be threatening was that, unlike the cycles of sun, moon, and seasons, the female cycle stopped. Men could fool themselves about growing old. They could tell themselves, and often young women also told them, that they were still young. They could shut their eyes to the passage of time. But a woman who had stopped menstruating was a living reminder that the passage of time is irrevocable, and that there is no restoration. This was most probably one reason why widows were burned in India, and witches in Europe and in America.

To be sure, not all widows and witches were old, but most widows are; and why wait? If the widow was still very young, one may have suspected that she was responsible for her husband's death. Perhaps it was just as well for her to have nothing to gain from it. The worse male Hindus treated their wives, the more they may well have felt that it was good if their wives knew that when their husbands died, they would die, too.

As for young witches, they were supposed to have been led astray by old witches. And why were old women so often believed to know magic? Perhaps their original magical trick was that while other cycles of nature continued endlessly and gave men the feeling that in the long run nothing really changes, women defied this pattern of nature.

The women's movement will have to address itself to these problems. The battle which has attracted the most attention is that of young women who insist on not being regarded as mere sex objects, drudges, or mothers. Many have decided not to have children. But all their victories will be hollow if they do not manage to change the prevalent attitudes toward the old.

In the past, ever growing numbers of women have dealt with this problem by not facing up to time. They have tried to hide time the way undertakers hide death, with cosmetics. They have claimed to be younger than they were or refused to tell their age. Time was their archenemy. They did not just fancy that; in the cultures in which they lived, it was.

48

Women may soon rally to the cry, "Old is beautiful!" Of course, not everything old is beautiful, any more than everything black, or everything white, or everything young. But the notion that old means ugly is every bit as harmful as the prejudice that black is ugly. In one way it is even more pernicious.

The notion that only what is new and young is beautiful poisons our relationship to the past and to our own future. It keeps us from understanding our roots and the greatest works of our culture and other cultures. It also makes us dread what lies ahead of us and leads many to shirk reality.

A large part of this book deals with works of art. Many people have a very limited interest in art. Some of the text deals with ancient Israel, India, and Greece. Most people have no intense concern with antiquity. But the central problem of the effects of time concerns all of us, even if millions refuse to think about it. It is not easy to get at this problem.

The philosophers who have written on time are not much help here. Almost all of them have ignored life at the limits and have done their work in the eye of life's hurricane.

My experience of time owes more to Rembrandt than to Plato, and more to the Hebrew Bible than to Kant. It has also been shaped by the contemplation of sculptures and ruins and of alarming "restorations." To Plato and Kant it never occurred that our experience of time could be shaped by looking at works of art, tree bark, erosion, and sunsets, or by reading stories like those of Jacob, Samson, or David.

To get modern readers to look at time in a new way, it would never do to follow the paths of the old philosophers. I have tried a new way, placing our experience of time in a temporal context, showing how it has historical roots, and above all offering pictures that are no less important than the text.

Of course, "Old is beautiful" is as paradoxical as "Time is an artist." What meets the eye is the opposite. Yet photographs show how both claims are true. Still they do not tell all, and it may be objected that this aesthetic approach is overly optimistic. My approach, however, is not merely aesthetic. It does not concentrate on surfaces while ignoring oppression, suffering, and death. On the contrary, this trilogy begins with *Life at the Limits*, and *Time Is an Artist* deals centrally with man's lot.

Time is an artist. But an artist is not only an artist. Old is beautiful. But old is not only beautiful. As long as we fail to see the artistry of time and the beauty of age, we are far from understanding man's lot. But to understand it more fully, we still need to ask: What is man?

BIBLIOGRAPHY

Angelus Silesius. *Sämtliche poetische Werke. In drei Bänden.* Ed. Hans Ludwig Held. München, Allgemeine Verlagsanstalt, 2nd rev. ed. 1924. The translation is from Kaufmann, *Twenty-five German Poets.*

Baeck, Leo. "Two World Views Compared" in *The Pharisees and Other Essays.* New York, Schocken, 1947. The original German version was first pulished in 1923 in *Der Leuchter,* iv, pp. 117–41, and was reprinted as the first essay in Baeck's *Wege im Judentum,* Berlin, Schocken, 1933, under the title "Vollendung und Spannung."

Coomaraswamy, Ananda K. *History of Indian and Indonesian Art.* New York, Dover Publications, Inc. 1965. First published in 1927.

Cornford, Francis M. *Thucydides Mythistoricus.* London, E. Arnold, 1907. Philadelphia, University of Pennsylvania Press, 1971 (same pagination).

Genesis. B. Jacob, *Das Erste Buch der Tora: Genesis, übersetzt und erklärt.* Berlin, Schocken, 1934. 1055 large pages. The most impressive modern commentary by an orthodox rabbi. E. A. Speiser, *Genesis: Introduction, Translation, and Notes.* Garden City, N.Y., Doubleday, 1964. The first volume of The Anchor Bible which embodies the most recent scholarship. See also Lowenthal, below.

Goldscheider, Ludwig. *Fünfhundert Selbstporträts von der Antike bis zur Gegenwart.* Phaidon-Verlag, Wien, 1936.

Gombrich, E. H. "Dark Varnishes: Variations on a Theme from Pliny," *Burlington Magazine,* 1962, pp. 51–55.

————. "Controversial Methods and Methods of Controversy," *Burlington Magazine,* 1963, pp. 90–93.

Goya, Don Francisco Zapater y Gómez, *Colección de Cuatrocientas de Cuadros, Dibujos y Aguafuertes de Don Francisco de Goya Precedidos de un Epistolario del gran pintor y de las Noticias Biograficas.* Madrid, Editorial "Saturnino Calleja" S.A. 1924. Goya's letter of 1801, quoted in section 37, was very graciously translated by Professor James E. Irby especially for this volume.

Huxtable, Ada Louise. "Alms for the Acropolis," *The New York Times,* January 17, 1977, Editorial page.

Kaufmann, Walter. *Critique of Religion and Philosophy.* New York, Harper, 1958. Text and section numbers unchanged in subsequent editions.

————. *Cain and Other Poems.* New York, Doubleday, 1962. 3rd enlarged edition, New York, New American Library, 1975. "Jacob" first appeared in *Cain;* "Balaam" and "Old" have not appeared in print before.

————. *Tragedy and Philosophy.* New York, Doubleday, 1968. Section numbers unchanged in subsequent editions.

————. *Twenty-five German Poets: A Bilingual Collection,* edited, translated, and introduced by Walter Kaufmann. New York, W. W. Norton, 1975.

————. *Man's Lot: A Trilogy.* Photographs and text. New York, Reader's Digest Press, 1978. Distributed by McGraw-Hill.

————. *Life At the Limits.* Photographs and text. New York, Reader's Digest Press, 1978. Distributed by McGraw-Hill.

————. *What Is Man?* Photographs and text. New York, Reader's Digest Press, 1978. Distributed by McGraw-Hill.

Kohn, Hans. *The Idea of Nationalism.* New York, Macmillan, 1948.

Kurz, Otto. "Varnishes, Tinted Varnishes, and Patina," *Burlington Magazine,* 1962, pp. 56–59.

————. "Time the Painter," *Burlington Magazine,* 1963, pp. 94–97.

Lowenthal, Eric I. *The Joseph Narrative in Genesis.* New York, Ktav Publishing House, Inc., 1973.

Meyer, Eduard. *Geschichte des Altertums:* vol. I. 1: *Einleitung. Elemente der Anthropologie.* 3rd ed. Stuttgart und Berlin, J. G. Cotta'sche Buchhandlung Nachfolger, 1910. The translation is mine.

Michelangelo. Charles De Tolnay, *Michelangelo.* 5 vols. Princeton, Princeton University Press, 1943ff. 2nd printing, with foreword, 1969–71. See also Erwin Panofsky, "The Neoplatonic Movement and Michelangelo" in *Studies in Iconology.* New York, Oxford University Press, 1939.

Pritchard, James B., ed. *Ancient Near Eastern Texts Relating to the Old Testament.* Princeton, Princeton University Press, 1950. 2nd edition, corrected and enlarged, 1955.

Rilke, Rainer Maria. *Sämtliche Werke,* vol. I. Insel Verlag, 1926. The translation is from Kaufmann, *Twenty-five German Poets.*

Singh, Khushwant. "We Sell Them Dreams: The Indian Cinema" in *India Without Humbug.* Bombay, India Book House, 1977.

With, Karl. *Java: Brahmanische, Buddhistische und Eigenlebige Architektur und Plastik auf Java: 165 Abbildungen und 13 Grundrisse.* Hagen i. W., Folkwang Verlag, 1920.

51

52

53

54

60

61

69

70